Model-Based Tools for Pharmaceutical Manufacturing Processes

Model-Based Tools for Pharmaceutical Manufacturing Processes

Special Issue Editors

Krist V. Gernaey
René Schenkendorf
Dimitrios I. Gerogiorgis
Seyed Soheil Mansouri

MDPI • Basel • Beijing • Wuhan • Barcelona • Belgrade • Manchester • Tokyo • Cluj • Tianjin

Special Issue Editors
Krist V. Gernaey
Technical University of Denmark
Denmark

René Schenkendorf
TU Braunschweig
Germany

Dimitrios I. Gerogiorgis
University of Edinburgh
UK

Seyed Soheil Mansouri
Technical University of Denmark
Denmark

Editorial Office
MDPI
St. Alban-Anlage 66
4052 Basel, Switzerland

This is a reprint of articles from the Special Issue published online in the open access journal *Processes* (ISSN 2227-9717) (available at: http://www.mdpi.com).

For citation purposes, cite each article independently as indicated on the article page online and as indicated below:

LastName, A.A.; LastName, B.B.; LastName, C.C. Article Title. *Journal Name* **Year**, *Article Number*, Page Range.

ISBN 978-3-03928-424-5 (Pbk)
ISBN 978-3-03928-425-2 (PDF)

Contents

About the Special Issue Editors vii

René Schenkendorf, Dimitrios I. Gerogiorgis, Seyed Soheil Mansouri and Krist V. Gernaey
Model-Based Tools for Pharmaceutical Manufacturing Processes
Reprinted from: *Processes* **2020**, *8*, 49, doi:10.3390/pr8010049 1

Robert T. Giessmann, Niels Krausch, Felix Kaspar, Mariano Nicolas Cruz Bournazou, Anke Wagner, Peter Neubauer and Matthias Gimpel
Dynamic Modelling of Phosphorolytic Cleavage Catalyzed by Pyrimidine-Nucleoside Phosphorylase
Reprinted from: *Processes* **2019**, *7*, 380, doi:10.3390/pr7060380 5

Andrew B. Cuthbertson, Alistair D. Rodman, Samir Diab and Dimitrios I. Gerogiorgis
Dynamic Modelling and Optimisation of the Batch Enzymatic Synthesis of Amoxicillin
Reprinted from: *Processes* **2019**, *7*, 318, doi:10.3390/pr7060318 19

Satyajeet Bhonsale, Carlos André Muñoz López and Jan Van Impe
Global Sensitivity Analysis of aSpray Drying Process
Reprinted from: *Processes* **2019**, *7*, 562, doi:10.3390/pr7090562 37

Peter Toson, Pankaj Doshi and Dalibor Jajcevic
Explicit Residence Time Distribution of a Generalised Cascade of Continuous Stirred Tank Reactors for a Description of Short Recirculation Time (Bypassing)
Reprinted from: *Processes* **2019**, *7*, 615, doi:10.3390/pr7090615 61

Nirupaplava Metta, Michael Ghijs, Elisabeth Schäfer, Ashish Kumar, Philippe Cappuyns, Ivo Van Assche, Ravendra Singh, Rohit Ramachandran, Thomas De Beer, Marianthi Ierapetritou and Ingmar Nopens
Dynamic Flowsheet Model Development and Sensitivity Analysis of a Continuous Pharmaceutical Tablet Manufacturing Process Using the Wet Granulation Route
Reprinted from: *Processes* **2019**, *7*, 234, doi:10.3390/pr7040234 75

Daniel Laky, Shu Xu, Jose S. Rodriguez, Shankar Vaidyaraman, Salvador García Muñoz and Carl Laird
An Optimization-Based Framework to Define the Probabilistic Design Space of Pharmaceutical Processes with Model Uncertainty
Reprinted from: *Processes* **2019**, *7*, 96, doi:10.3390/pr7020096 111

Xiangzhong Xie and René Schenkendorf
Robust Process Design in Pharmaceutical Manufacturing under Batch-to-Batch Variation
Reprinted from: *Processes* **2019**, *7*, 509, doi:10.3390/pr7080509 131

Haruku Shirahata, Sara Badr, Yuki Shinno, Shuta Hagimori and Hirokazu Sugiyama
Online Decision-Support Tool "TECHoice" for the Equipment Technology Choice in Sterile Filling Processes of Biopharmaceuticals
Reprinted from: *Processes* **2019**, *7*, 448, doi:10.3390/pr7070448 145

Christos Varsakelis, Sandrine Dessoy, Moritz von Stosch and Alexander Pysik
Show Me the Money! Process Modeling in Pharma from the Investor's Point of View
Reprinted from: *Processes* **2019**, *7*, 596, doi:10.3390/pr7090596 165

About the Special Issue Editors

Krist V. Gernaey has a Ph.D. (1997) from Ghent University (Belgium). His Ph.D. research focused on the monitoring of wastewater systems. He held postdoc positions at Ghent University, École Polytechnique de Montreal, the Technical University of Denmark and Lund University from 1998 to 2005, performing research on modeling, simulation and control of wastewater systems. He has been an associate professor at the Department of Chemical and Biochemical Engineering at the Technical University of Denmark (DTU) since 2005 and professor in industrial fermentation technology ("The Novozymes professor") since 2013. He is the head of the Process and Systems Engineering Center (PROSYS) since 2014. Further, he has been the CEO of Bioscavenge ApS since 2017, a company with a focus on resource recovery from waste streams from the biotech industry. Prof. Gernaey's current research is focused on large-scale fermentation, novel sensor technologies, mathematical modeling, mass transfer issues across scales, process simulation, and digital twins.

René Schenkendorf received his Dipl.-Ing. and Dr.-Ing. in Engineering Cybernetics from the Otto-von-Guericke-University Magdeburg, Germany in 2007 and 2014, respectively. From 2007 to 2012, he was a Ph.D. student at the Max Planck Institute for Dynamics of Complex Technical Systems in Magdeburg, Germany. From 2013 to 2016, he was with the German Aerospace Center at the Institute of Transportation Systems in Braunschweig, Germany. Since 2016, he has been the head of the Process Systems Engineering group at the Institute of Energy and Process Systems Engineering at the Technische Universität Braunschweig, Germany. His current research interests include process systems engineering, sensitivity/uncertainty analysis, and model-based optimal experimental design in the field of pharmaceutical manufacturing.

Dimitrios I. Gerogiorgis is a Senior Lecturer in Chemical Engineering and Director of the MSc in Advanced Chemical Engineering of the University of Edinburgh, focusing on process systems modeling, design, and optimization. He holds a Diploma and Ph.D. in Chem. Eng., an MSc in Elec. and Comput. Eng., a degree in Higher Education and a Diploma in Translation. He has been honored with a Royal Academy of Engineering (RAEng) Industrial Fellowship (2017), a High Commendation for the IChemE Global Food and Drink Award (2017), and the Academy of Athens "Loukas Mousoulos" Publication Research Excellence Prize (2015). His course on Oil & Gas Sys. Eng. (with Atkins) has also been shortlisted for the IChemE Global Education & Training Award (2015).

Seyed Soheil Mansouri is an Assistant Professor at the Department of Chemical and Biochemical Engineering at the Technical University of Denmark (DTU) since February 2018 and affiliate faculty at the Sino-Danish Center for Education and Research in Beijing, China. He received his Ph.D. (2016) and MSc (2013) in chemical and biochemical engineering, both from DTU. His current research is primarily focused on developing systematic methods and tools for synthesis, design, and control/optimization of chemical and bio-pharmaceutical processes with an aim to achieve more sustainable production and consumption. He is a senior member of the American Institute of Chemical Engineers (AIChE) and a Danish delegate to the Computer Aided Process Engineering (CAPE) Working Party of the European Federation of Chemical Engineering (EFCE).

processes

MDPI

Editorial

Model-Based Tools for Pharmaceutical Manufacturing Processes

René Schenkendorf [1,2,*], Dimitrios I. Gerogiorgis [3], Seyed Soheil Mansouri [4] and Krist V. Gernaey [4]

1 Institute of Energy and Process Systems Engineering, Technische Universität Braunschweig, Franz-Liszt-Straße 35, 38106 Braunschweig, Germany
2 Center of Pharmaceutical Engineering (PVZ), Technische Universität Braunschweig, Franz-Liszt-Straße 35a, 38106 Braunschweig, Germany
3 Institute for Materials and Processes (IMP), School of Engineering, University of Edinburgh, The King's Buildings, Edinburgh EH9 3FB, UK; d.gerogiorgis@ed.ac.uk
4 Department of Chemical and Biochemical Engineering, Technical University of Denmark, Building 229, 2800 Kongens Lyngby, Denmark; seso@kt.dtu.dk (S.S.M.); kvg@kt.dtu.dk (K.V.G.)
* Correspondence: r.schenkendorf@tu-braunschweig.de

Received: 5 November 2019; Accepted: 27 December 2019; Published: 1 January 2020

Active pharmaceutical ingredients (APIs) are highly valuable, highly sensitive products resulting from production processes with strict quality control specifications and regulations that are required for the safety of patients. To ensure a profitable and growing pharmaceutical industry of significant societal benefits and low environmental footprint, model-based tools are fundamental to advancing the basic understanding, design, and optimization of pharmaceutical manufacturing processes in accordance with the United Nations "2030 sustainable development goals". Process analysis principles, for instance, provide a better understanding of underlying pharmaceutical manufacturing mechanisms. Model-based process design concepts facilitate the identification of optimal production and purification pathways and configurations. Process monitoring and control strategies ensure low life-cycle costs and provide new insights into critical failure modes and drug quality control issues.

The foregoing model-based concepts, and combinations of them, are key to exploring the full potential of innovative, highly effective pharmaceutical manufacturing processes. These are some of the grand challenges that can be tackled by process systems engineering (PSE), and they have been catalyzed by an unprecedented advent of established methodologies and algorithmic tools that are either available via open access environments or incorporated in commercial software/databases for a plethora of purposes (thermodynamic and solubility modeling, fluid phase equilibria, complex mixture thermophysical/mechanical property estimation, plant-wide simulation, optimization and cost estimation). The respective advances achieved using such a diversity of enabling computational technologies exemplify the Quality-by-Design (QbD) vision and its translation into tangible artefacts and policies, illustrating how academia and industry respond to contemporary challenges for high-quality, more affordable healthcare.

This Special Issue on "Model-Based Tools for Pharmaceutical Manufacturing Processes" intends to curate novel advances in the development and application of model-based tools to address ever-present challenges of the traditional pharmaceutical manufacturing practice as well as new trends. As summarized below, the Special Issue provides a collection of nine papers on original advances in the model-based process unit, system-level, QbD under uncertainty, and decision-making applications of pharmaceutical manufacturing processes. The Special Issue is available online at https://www.mdpi.com/journal/processes/special_issues/pharmaceutical_processes.

1. Process Unit Studies

Before complex pharmaceutical manufacturing processes can be simulated holistically, dedicated unit operations have to be studied first, including upstream and downstream processes. Starting with the synthesis of APIs and relevant intermediates, enzymatic syntheses are of particular interest as a greener, more economical, and efficient viable alternative to chemocatalytic processes. For instance, pyrimidine-nucleoside phosphorylases are highly versatile enzymes used for the production of pharmaceutically relevant intermediates. In "Dynamic Modelling of Phosphorolytic Cleavage Catalyzed by Pyrimidine-Nucleoside Phosphorylase" [1], the conversion of deoxythymidine and phosphate to deoxyribose-1-phosphate and thymine by a thermophilic pyrimidine-nucleoside phosphorylase from *Geobacillus thermoglucosidasius* was modeled in detail and validated experimentally including UV/Vis spectroscopy data. The resulting dynamic model might be used to identify optimal operating conditions of the enzymatic synthesis process, and can be extended to multi-enzyme reactions too.

Moreover, antibiotics are an essential group of biologics, and thus of interest in pharmaceutical manufacturing. In "Dynamic Modelling and Optimisation of the Batch Enzymatic Synthesis of Amoxicillin" [2], the batch enzymatic synthesis of the antibiotic amoxicillin, listed as a World Health Organization (WHO) "Essential Medicine", was modeled and optimized. While including non-isothermal kinetics, the authors identified an optimal temperature profile that ensures high product quality at minimum feedstock consumption.

In addition to synthesis problems, modeling of downstream processes has attracted much interest in the last few decades. For instance, spray drying is a basic unit operation in pharmaceutical manufacturing. In "Global Sensitivity Analysis of a Spray Drying Process" [3], a sensitivity analysis study of a spray drying process is discussed. To quantify the impact of different but interacting process parameters, a model-based global sensitivity analysis with a low computational cost was implemented, contributing to QbD and the identification of critical process parameters. These essential parameters of the process might be relevant for the development of future control strategies that can result in significant robustness for the spray drying process.

2. System-Level Studies

Next, based on determined process unit models, system-level studies are crucial for a detailed understanding of pharmaceutical manufacturing processes. The interaction between process units, the identification of critical process parameters, and their impact on critical quality attributes of pharmaceutical products are of key interest at the system level. For instance, when modeling the flow of material in a continuous process of several unit operations (e.g., blending, granulation, and tableting), the study of residence time distributions is the tool of choice. In "Explicit Residence Time Distribution of a Generalised Cascade of Continuous Stirred Tank Reactors for a Description of Short Recirculation Time (Bypassing)" [4], the so-called tanks-in-series model was generalized to a cascade of n continuous stirred tank reactors with non-integer non-negative n. Therefore, the model can describe short recirculation times (bypassing) without the need for complex reactor networks. When part of a reactor network, the proposed model can be used to predict the response to upstream setpoint changes and process fluctuations, i.e., providing insights at the system level.

The relevance of model-based studies of process-wide manufacturing lines is highlighted in "Dynamic Flowsheet Model Development and Sensitivity Analysis of a Continuous Pharmaceutical Tablet Manufacturing Process Using the Wet Granulation Route" [5]. In this study, the authors implemented a dynamic flowsheet model of the ConsiGma^{TM}-25 line for continuous tablet manufacturing, including determined models of various unit operations, i.e., feeders, blenders, a twin-screw wet granulator, a fluidized bed dryer, a mill, and a tablet press. Based on the developed dynamic flowsheet model, the liquid feed rate to the granulator, the air temperature, and the drying time in the dryer were identified via global sensitivity analysis methods as critical process parameters that affect the tablet properties most.

3. Studies Under Uncertainty

QbD, an essential paradigm in pharmaceutical manufacturing, benefits from mathematical models. Model imperfections, however, have to be considered seriously; that is, uncertainty quantification and analysis are mandatory in model-based studies. This is particularly true when making use of mathematical models to study the design space of manufacturing processes. In "An Optimization-Based Framework to Define the Probabilistic Design Space of Pharmaceutical Processes with Model Uncertainty" [6], the authors introduced two algorithms to analyze the design space under uncertainties at low computational costs. The usefulness of the proposed probabilistic design space implementations was benchmarked with pharmaceutical manufacturing problems, including the Michael addition reaction as an industrial relevant case study.

In addition to uncertain model parameters and kinetics, batch-to-batch variations cause severe difficulties in pharmaceutical manufacturing, affecting drug quality, clinical studies, and therapeutics in equal measure. The joint effect of model imperfection and batch-to-batch variation is addressed in "Robust Process Design in Pharmaceutical Manufacturing under Batch-to-Batch Variation" [7]. Considering a freeze-drying process, the authors used an efficient model-based concept to predict optimal shelf temperature and chamber pressure profiles under batch-to-batch variation.

4. Decision-Making Studies

In addition to process analysis and optimization in pharmaceutical manufacturing, mathematical models can support the decision-making process in identifying the best manufacturing concepts in terms of reduced capital and operating costs. For instance, the best choice between pharmaceutical manufacturing process alternatives is challenging and benefits considerably from algorithms and decision-making tools. In "Online Decision-Support Tool "TECHoice" for the Equipment Technology Choice in Sterile Filling Processes of Biopharmaceuticals" [8], the authors proposed a model-based tool to support users in choosing their preferred technology according to their input of specific drug production scenarios. The usefulness of the prototype tool was demonstrated successfully with the study of equipment technologies in the sterile filling of biopharmaceutical manufacturing processes.

Modeling and simulation are a central part of research and development activities in the pharmaceutical industry, but the evaluation of modeling and simulation return on investments is difficult to quantify in advance. In "Show Me the Money! Process Modeling in Pharma from the Investor's Point of View" [9], the authors provide an algorithmic methodology that allows for the development of detailed business studies. They discuss an easy-to-use methodology that can help an investor evaluate an investment in modeling and simulation systematically.

The present Special Issue on "Model-Based Tools for Pharmaceutical Manufacturing Processes" and several more on adjacent topics which have either appeared or will be featured in *Processes* (but also in journals of similar scope and mission) signify the rapidly expanding importance of this research field towards securing sophisticated healthcare solutions and improving accessibility to medication for the ever-increasing and ageing global population. Publishing the fruits of academic, industrial, and collaborative efforts to this end should serve as inspiration for new challenges to set and solutions to achieve; our fervent hope is hence that PSE contributions will remain front and center in this quest.

References

1. Giessmann, R.T.; Krausch, N.; Kaspar, F.; Cruz Bournazou, M.N.; Wagner, A.; Neubauer, P.; Gimpel, M. Dynamic Modelling of Phosphorolytic Cleavage Catalyzed by Pyrimidine-Nucleoside Phosphorylase. *Processes* **2019**, *7*, 380. [CrossRef]
2. Cuthbertson, A.B.; Rodman, A.D.; Diab, S.; Gerogiorgis, D.I. Dynamic Modelling and Optimisation of the Batch Enzymatic Synthesis of Amoxicillin. *Processes* **2019**, *7*, 318. [CrossRef]
3. Bhonsale, S.; Muñoz López, C.A.; Van Impe, J. Global Sensitivity Analysis of a Spray Drying Process. *Processes* **2019**, *7*, 562. [CrossRef]

4. Toson, P.; Doshi, P.; Jajcevic, D. Explicit Residence Time Distribution of a Generalised Cascade of Continuous Stirred Tank Reactors for a Description of Short Recirculation Time (Bypassing). *Processes* **2019**, *7*, 615. [CrossRef]
5. Metta, N.; Ghijs, M.; Schäfer, E.; Kumar, A.; Cappuyns, P.; Assche, I.V.; Singh, R.; Ramachandran, R.; De Beer, T.; Ierapetritou, M.; et al. Dynamic Flowsheet Model Development and Sensitivity Analysis of a Continuous Pharmaceutical Tablet Manufacturing Process Using the Wet Granulation Route. *Processes* **2019**, *7*, 234. [CrossRef]
6. Laky, D.; Xu, S.; Rodriguez, J.; Vaidyaraman, S.; García Muñoz, S.; Laird, C. An Optimization-Based Framework to Define the Probabilistic Design Space of Pharmaceutical Processes with Model Uncertainty. *Processes* **2019**, *7*, 96. [CrossRef]
7. Xie, X.; Schenkendorf, R. Robust Process Design in Pharmaceutical Manufacturing under Batch-to-Batch Variation. *Processes* **2019**, *7*, 509. [CrossRef]
8. Shirahata, H.; Badr, S.; Shinno, Y.; Hagimori, S.; Sugiyama, H. Online Decision-Support Tool "TECHoice" for the Equipment Technology Choice in Sterile Filling Processes of Biopharmaceuticals. *Processes* **2019**, *7*, 448. [CrossRef]
9. Varsakelis, C.; Dessoy, S.; von Stosch, M.; Pysik, A. Show Me the Money! Process Modeling in Pharma from the Investor's Point of View. *Processes* **2019**, *7*, 596. [CrossRef]

processes

MDPI

Article

Dynamic Modelling of Phosphorolytic Cleavage Catalyzed by Pyrimidine-Nucleoside Phosphorylase

Robert T. Giessmann [1,*,†], Niels Krausch [1,†], Felix Kaspar [1], Mariano Nicolas Cruz Bournazou [2,3], Anke Wagner [1,4], Peter Neubauer [1] and Matthias Gimpel [1]

1 Laboratory of Bioprocess Engineering, Department of Biotechnology, Technische Universität Berlin, Ackerstr. 76, ACK24, D-13355 Berlin, Germany; n.krausch@tu-berlin.de (N.K.); f.kaspar@tu-braunschweig.de (F.K.); anke.wagner@tu-berlin.de (A.W.); peter.neubauer@tu-berlin.de (P.N.); matthias.gimpel@tu-berlin.de (M.G.)
2 Institute of Chemical and Bioengineering, Department of Chemistry and Applied Biosciences, ETH Zürich, Vladimir-Prelog-Weg 1, 8093 Zurich, Switzerland; n.cruz@datahow.ch
3 DataHow AG, Vladimir-Prelog-Weg 1, 8093 Zurich, Switzerland
4 BioNukleo GmbH, Ackerst. 76, D-13355 Berlin, Germany
* Correspondence: r.giessmann@tu-berlin.de
† Authors contributed equally to this paper.

Received: 27 May 2019; Accepted: 13 June 2019; Published: 19 June 2019

Abstract: Pyrimidine-nucleoside phosphorylases (Py-NPases) have a significant potential to contribute to the economic and ecological production of modified nucleosides. These can be produced via pentose-1-phosphates, an interesting but mostly labile and expensive precursor. Thus far, no dynamic model exists for the production process of pentose-1-phosphates, which involves the equilibrium state of the Py-NPase catalyzed reversible reaction. Previously developed enzymological models are based on the understanding of the structural principles of the enzyme and focus on the description of initial rates only. The model generation is further complicated, as Py-NPases accept two substrates which they convert to two products. To create a well-balanced model from accurate experimental data, we utilized an improved high-throughput spectroscopic assay to monitor reactions over the whole time course until equilibrium was reached. We examined the conversion of deoxythymidine and phosphate to deoxyribose-1-phosphate and thymine by a thermophilic Py-NPase from *Geobacillus thermoglucosidasius*. The developed process model described the reactant concentrations in excellent agreement with the experimental data. Our model is built from ordinary differential equations and structured in such a way that integration with other models is possible in the future. These could be the kinetics of other enzymes for enzymatic cascade reactions or reactor descriptions to generate integrated process models.

Keywords: enzymatic reaction; reversible reaction; dynamic modelling; pyrimidine-nucleoside phosphorylase; spectroscopic assay; process kinetics; ODE model

1. Introduction

Pyrimidine-nucleoside phosphorylases (Py-NPases) are highly versatile enzymes used for the production of pharmaceutically relevant nucleoside derivatives and pentose-1-phosphates. Generally, nucleoside phosphorylases catalyze, in the presence of phosphate, the reversible conversion of a nucleoside to the corresponding pentose-1-phosphate and nucleobase (Figure 1). Due to the low yields of modified nucleosides or pentose-1-phosphates via conventional synthetic chemistry, nucleoside phosphorylases have become attractive tools in their biocatalytic preparation [1–3]. Recently, thermophilic Py-NPases have attracted increased interest, as they combine several favorable properties, such as long shelf life due to their thermal stability, an excellent tolerance towards harsh reaction conditions, high turnover rates, and a broad substrate spectrum [4,5].

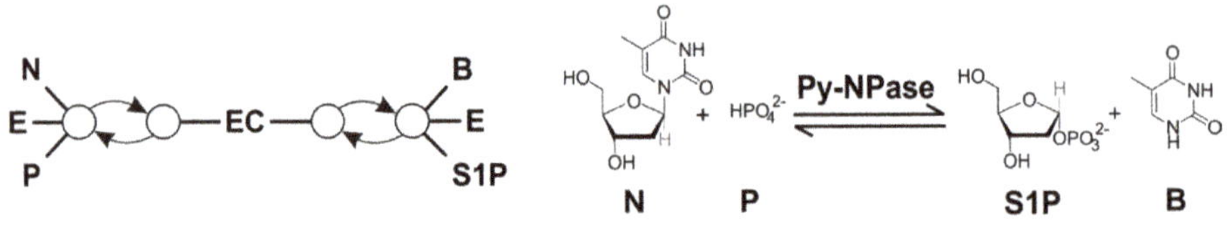

(a) (b)

Figure 1. Schematic and chemical illustration of an enzymatic nucleoside phosphorylation. (**a**) Schematic drawing of the proposed mechanics for an enzymatic nucleoside phosphorylation reaction as basis for the generation of the differential–dynamical model. Enzyme (E), nucleoside (N), and phosphate (P) react in a three-particle collision towards the enzyme complex (EC), which decays without other intermediates into enzyme, pentose-1-phosphate (S1P), and free nucleobase (B). Both reactions can occur in the other direction, as well; (**b**) chemical structures of an enzymatic phosphorylation using the example of the enzyme pyrimidine-nucleoside phosphorylase (Py-NPase; E) catalyzed reaction of the nucleoside deoxythymidine (N) and ortho-phosphate (P) to the free nucleobase thymine (B) and deoxyribose-1-phosphate (S1P).

However, their industrial use is hampered by a lack of models which integrate the understanding of their behavior in enzymatic reactions over the full time course towards the reaction's dynamic equilibrium. Previous research has focused on either: (1) Integrated processes, mainly with transglycosylation and/or product removal reactions, which renders modelling of the complete process unfeasible because of its complexity; or (2) Michaelis–Menten conditions, i.e., reactions in which one of the substrates (typically phosphate) is present in excess over the other substrate, and only initial rates are measured (reviewed in [6]). This is because the Michaelis–Menten assumptions are only fulfilled in the quasi-linear range of conversion at the very start of the enzymatic reaction. Only in this time frame one can observe a constant conversion rate. Invariably, this only allows for the investigation of the dependence of the initial rate of the reaction on the concentration of a substrate and does not permit the evaluation of the whole time-course [6].

In industrial applications, the stoichiometric and quantitative conversion of substrates is highly anticipated. These requirements are only met when the reaction approaches its thermodynamic equilibrium, hence giving maximum product yield. Counteracting the accessibility of deoxyribose-1-phosphate is the fact that the equilibrium for nucleoside phosphorylation reactions is strongly in favor of the substrates (K_{eq} = 0.03–0.10 for pyrimidines [7,8], and K_{eq} = 0.01–0.02 for purines [9,10]). To increase the concentration of desired products, it is therefore necessary to push the equilibrium, e.g., by increasing the phosphate concentration. Despite the clear need for a Py-NPase model describing those industrially relevant conditions, there has been no report of a suitable model so far.

Models of ordinary differential equations (ODEs) derived from elementary reaction steps and from law of mass action kinetics ("differential–dynamical models") present an attractive solution to many biotechnological problems. Their modularity allows for the combination of models of different scales, such as the progression of an enzyme reaction with a substrate feeding profile. Differential–dynamical models have been used to describe, for example, enzymatic cellulose hydrolysis (reviewed in [11]), the production of enantiopure amines from a racemic mixture [12], the continuous production of lactobionic acid from lactose [13], or symmetric two-educts/one-product carboligations [14]. The rate laws of differential–dynamical models are usually derived from an underlying mechanical model. This enables chemical reaction engineering across different conditions and scales [15]. The ultimate promise of differential–dynamical models is the model-based design of dynamic experiments [16], which are favorable for biotechnological applications [17] and allow the in silico predictability of economic production processes [18], even for processes where the experimental information is scarce [19].

In this work, we present experimental data deduced from the reaction monitoring of small-scale Py-NPase reactions via a UV/Vis spectroscopy-based assay. Subsequently, we report the development of a differential–dynamical model for the Py-NPase-mediated biocatalytic preparation of deoxyribose-1-phosphate from thymidine.

2. Materials and Methods

2.1. Materials

All chemicals used in this study were of analytical grade and used without further purification. The water used in all solutions was deionized to 18.2 MΩ·cm with a water purification system from Werner. Deoxythymidine was purchased from Carbosynth. Thymine and phosphate (KH_2PO_4) were purchased from Sigma–Aldrich. Tris (2-Amino-2-(hydroxymethyl)propane-1,3-diol) was of buffer grade and purchased from Carl Roth.

Tris buffer was prepared as a 50 mM solution, and the pH was adjusted to 9.0 using 1 M HCl. Phosphate was prepared as a 1 M stock solution in 50 mM Tris buffer, and the pH was subsequently adjusted to 9.0 using 1 M NaOH. Deoxythymidine, and thymine stock solutions were prepared in different concentrations (ranging from 1 to 10 mM) by adding 50 mM of Tris buffer (of pH 9.0; the final pH of the prepared solution was found to be 9.0 as well) and treated with ultrasound to facilitate full dissolution.

The enzyme under investigation was a Py-NPase (EC 2.4.2.2, NCBI sequence accession number WP_041270053.1) from *Geobacillus thermoglucosidasius* (DSM No.: 2542). After IPTG-induced recombinant overexpression, the N-terminally His_6-tagged Py-NPase was purified from *E. coli* BL21 using Ni-NTA affinity chromatography, as described previously [20]. Purity was determined by SDS-PAGE analysis and found to be >90%. Subsequently, the enzyme was dialyzed against 2 mM potassium phosphate buffer, pH 7.0 (measured at 25 °C), and stored until use at +4 °C at a concentration of 3.69 mg/mL, as judged by NanoDrop analysis (calculated with 0.48 absorption units (AU) at 280 nm = 1 mg/mL). One unit (1 U) of enzyme activity was defined as the conversion of 1 μmol of deoxythymidine per minute in a 1 mL assay mixture of 2 mM deoxythymidine and 75 mM phosphate in 50 mM Tris buffer at a reaction temperature of 40 °C and at pH = 9.0 (measured at 25 °C), as determined by the method described later. The molecular weight of the enzyme was 47.6 kDa, as calculated from its amino acid sequence. The used enzyme preparation had an activity of approximately 0.46 U/mg.

UV/Vis transparent 96-well plates (UV-STAR F-Bottom #655801, purchased from Greiner Bio-One) were used to host the solutions for UV/Vis spectroscopy.

2.2. Experimental

Phosphate and deoxythymidine concentrations were varied in the range of 2–80 mM and 0.8–5 mM, respectively, in the assay mixture. The final enzyme concentration in the assay mixture was in the range of 12.5–50 μg/mL. This corresponds to an enzyme monomer concentration of 0.26–1.05 μM, as calculated from its molecular weight.

Reaction mixtures were prepared in 1.5 mL microreaction tubes. Appropriate amounts of phosphate and deoxythymidine stock solutions were added to an appropriate amount of the 50 mM Tris solution. All components were mixed by vortexing, and the microreaction tube preheated for at least 5 min in an Eppendorf ThermoStat Plus. Subsequently, an appropriate amount of enzyme stock solution was added to the tube, which was mixed by slight inversions. At given timepoints, a 60 μL sample was removed from the microreaction tube and injected immediately into 940 μL of a 0.2 M NaOH solution in a separate tube to stop the reaction and to dilute the sample simultaneously. After vortexing, 300 μL of the diluted mixture was transferred into UV/Vis transparent 96-well plates. When the concentration of UV/Vis absorbing compound, i.e., deoxythymidine or thymine, was varied, the sampling volume was adjusted as appropriate to give a constant final concentration of approximately 60 μM UV/Vis absorbing compounds in the alkaline dilutions to generate a UV/Vis absorption in the linear range, i.e., 0–1 absorption units (AU) at 260 nm. The ratio of substrate and product was determined by fitting the spectral 300/277 nm ratio (see below).

UV/Vis absorption spectra were recorded with a PowerWave HT or Synergy MX (BioTek Instruments, Bad Friedrichshall, Germany) in the range of 250–350 nm in 1 nm steps. Spectra were corrected for blanks, i.e., a 0.2 M NaOH solution, recorded within each set of measurements.

2.3. Spectroscopic Determination of Deoxythymidine/Thymine Ratio

The deoxythymidine/thymine ratio was determined with a spectrophotometric assay, modified from [21]. In an extension to previous versions of this assay, the spectra were normalized to the isosbestic point of deoxythymidine-thymine mixtures as suggested by [22], which we determined to be at 277 nm. This increased robustness against random dilution errors [23], as they commonly appear in high-throughput experimentation.

Briefly, the spectrum was first blank-corrected by subtracting a spectrum of 0.2 M NaOH, and was subsequently divided by its absorption at the isosbestic point to normalize the spectrum at this position to "1". Then, the normalized absorption at 300 nm was considered as a proxy of the deoxythymidine/thymine ratio.

Thus, the measured absorption ratio $Abs_{300/277} = Abs_{300}/Abs_{277}$ was fitted by a linear relationship without intercept:

$$\begin{aligned} Abs_{300/277}(\text{experimental}) = x \times Abs_{300/277}(\text{deoxythymidine}) + (1-x)\times \\ Abs_{300/277}(\text{thymine}), \end{aligned} \tag{1}$$

where x is the mole fraction of deoxythymidine in the mixture. From pure compound spectra, we determined $Abs_{300/277}$(deoxythymidine) = 0.005115 and $Abs_{300/277}$(thymine) = 0.772973.

The algorithms and data treatment functions were implemented in Python 2.7 [24] and Python 3.6 [25]. A snapshot of the software code and the data set used for this work is openly available on zenodo.org and in the Supplementary Material [26–29].

2.4. Modelling of the Py-NPase Catalyzed Reaction

The model was implemented as a system of ordinary differential equations in SymPy [30]. The system of equations was wrapped by python-sundials [31] and subsequently integrated by SUNDIALS-CVODE [32]. Parameter estimation was conducted via the lmfit interface [33]. The experimental data handling was performed by in-house Python software, which is equally available from the sources mentioned above.

2.4.1. Cost Functions

In the parameter estimation of the dynamic system (i.e., time courses of the reactions), a weighted-least squares cost function Z was used:

$$Z(k) = \sum_{i=1}^{Q} \frac{1}{\text{Var}(x_i)} \times (c(y_i) - c(x_i))^2, \tag{2}$$

where k is the parameter set used for calculation of the modelled concentrations; $\text{Var}(x_i)$ is the variance of the i-th data point; $c(y_i)$ is the modelled concentration of nucleoside for i-th data point; $c(x_i)$ is the nucleoside concentration as calculated from the experimentally determined mole fraction of deoxythymidine for i-th data point, multiplied with $c_{0(x_i)}$, i.e., the designed nucleoside concentration at $t = 0$; and Q is the total number of data points.

For the determination of weights, the 95% confidence interval of data points was set to 5 percentage points of the determined mole fraction as judged by inspection of calibration plots (Figure S1):

$$Var(x_i) = \left(\frac{\sqrt{\varepsilon}}{z_{0.975}} \times x_i\right) \times c_{0(x_i)}, \tag{3}$$

where $\varepsilon = 0.05$ gives the absolute error of the analysis method, and $z_{0.975} = 1.96$ gives the standard score to include 95% of values.

2.4.2. Definition of the Differential–Dynamical Model

A schematic visualization of the mechanical model is shown in Figure 1a, with specification into its chemical meaning in Figure 1b. The underlying mechanics of our differential–dynamical model at the process scale can also be represented indirectly by Scheme 1:

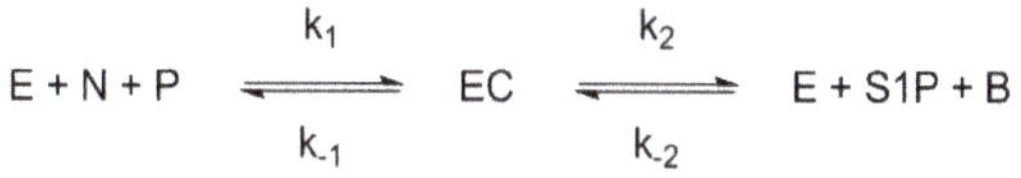

Scheme 1. Reaction equation of an enzymatic nucleoside phosphorylation. Enzyme (E), nucleoside (N), phosphate (P), enzyme complex (EC), pentose-1-phosphate (S1P), free nucleobase (B), reaction rate constants (k_1, k_{-1}, k_2, k_{-2}) as defined by Equations (7)–(10).

All steps indicated in the representation of the mechanics are considered elementary step reactions, and, applying law of mass action, the reaction rate equations are derived as the following system of ordinary differential equations:

$$\frac{d[\mathrm{N}]}{dt} = \frac{d[\mathrm{P}]}{dt} = -r_1 + r_{-1} \tag{4}$$

$$\frac{d[\mathrm{E}]}{dt} = -\frac{d[\mathrm{EC}]}{dt} = -r_1 + r_{-1} + r_2 - r_{-2} \tag{5}$$

$$\frac{d[\mathrm{S1P}]}{dt} = \frac{d[\mathrm{B}]}{dt} = +r_2 - r_{-2} \tag{6}$$

where [N] is the concentration of nucleoside (i.e., deoxythymidine), [P] is the concentration of phosphate, [E] is the concentration of free enzyme, [EC] is the concentration of enzyme complex, [S1P] is the concentration of pentose-1-phosphate (i.e., deoxyribose-1-phosphate), and [B] is the concentration of nucleobase (i.e., thymine), with the following rates:

$$r_1 = k_1 \times [\mathrm{E}] \times [\mathrm{N}] \times [\mathrm{P}] \tag{7}$$

$$r_{-1} = k_{-1} \times [\mathrm{EC}] \tag{8}$$

$$r_2 = k_2 \times [\mathrm{EC}] \tag{9}$$

$$r_{-2} = k_{-2} \times [\mathrm{E}] \times [\mathrm{S1P}] \times [\mathrm{B}] \tag{10}$$

3. Results

3.1. The Absorption Spectrum of Thymine but Not Deoxythymidine Changes at Alkaline Conditions

The evaluation of enzymatic deoxyribose-1-phosphate forming reactions requires the fast detection of substrates and products. The detection of nucleoside and its corresponding nucleobase by HPLC, and thus the indirect determination of pentose-1-phosphate, has been the standard method to date (e.g., [8,34]). However, it is very time-consuming and laborious and therefore not suitable for use in high-throughput screenings.

We intended to measure the deoxythymidine/thymine ratio by following wavelengths at regions where thymine absorbs at high pH, but deoxythymidine does not, based on an early report [21], and the more recent employment of an UV/Vis assay based on this principle [35]. These are wavelengths >290 nm [36–38]. To correct for varying path lengths which are commonly observed in high-throughput environments based on microtiter plates, and, thus, to make the assay more robust, we normalized the spectra to their isosbestic point, i.e., the point where no change in absorption is observed for any mixture of deoxythymidine and thymine.

To verify this concept experimentally, spectra of pure deoxythymidine, thymine, and mixtures of both were recorded after dilution in NaOH (Figure 2). We then calculated the composition of mixtures from the $Abs_{300/277}$ ratios as described in Materials and Methods. The composition of the full range of mixtures (0–100%, in 10% steps) could be estimated with high accuracy, and the absolute errors between the predicted and actual composition of the mixtures were approximately constant (Figure S1). With this high-throughput tool in hand, we pursued our investigation of a Py-NPase-catalyzed phosphorylation reaction and set out to describe our experimental data in a suitable model.

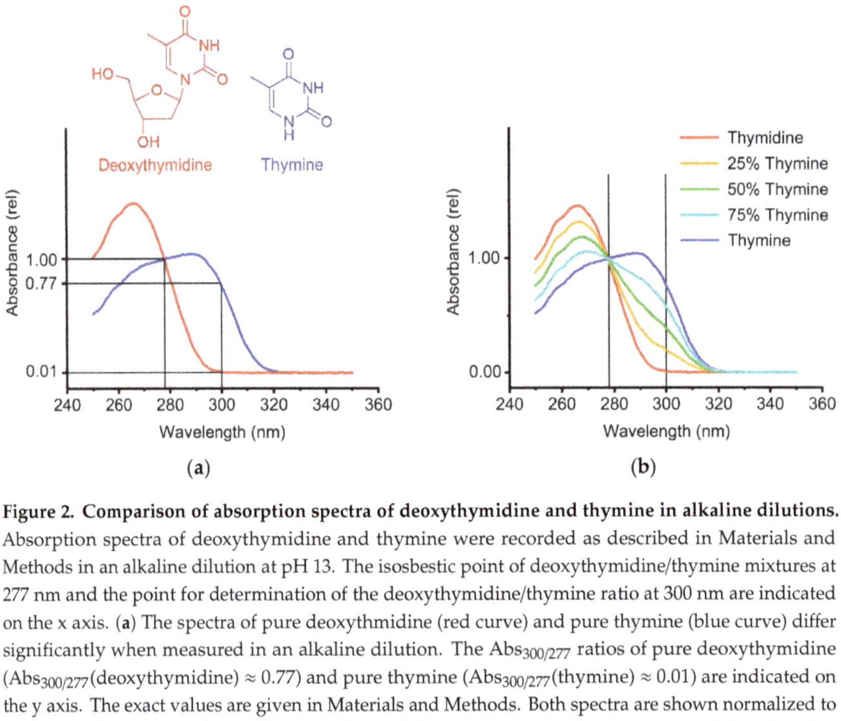

Figure 2. Comparison of absorption spectra of deoxythymidine and thymine in alkaline dilutions. Absorption spectra of deoxythymidine and thymine were recorded as described in Materials and Methods in an alkaline dilution at pH 13. The isosbestic point of deoxythymidine/thymine mixtures at 277 nm and the point for determination of the deoxythymidine/thymine ratio at 300 nm are indicated on the x axis. (**a**) The spectra of pure deoxythmidine (red curve) and pure thymine (blue curve) differ significantly when measured in an alkaline dilution. The $Abs_{300/277}$ ratios of pure deoxythymidine ($Abs_{300/277}$(deoxythymidine) $\approx$ 0.77) and pure thymine ($Abs_{300/277}$(thymine) $\approx$ 0.01) are indicated on the y axis. The exact values are given in Materials and Methods. Both spectra are shown normalized to the isosbestic point at 277 nm; (**b**) comparison of absorption spectra of pure deoxythmidine, thymine, and indicated mixtures, measured in an alkaline dilution. $Abs_{300/277}$ increases linearly with increasing thymine mole fraction (given as percentage; from red to blue).

3.2. *Model and Experimental Data Are in Excellent Agreement*

Py-NPase-catalyzed phosphorylic cleavage reactions are reversible reactions proceeding towards a dynamic equilibrium. Therefore, the reaction trajectory until equilibrium does not only depend on physical parameters, like temperature and pressure, but also on enzyme concentration, the concentration of substrates, or the presence of products. In order to investigate this enzymatic reaction under biotechnologically relevant conditions, we performed 48 experiments with varying concentrations of enzyme, nucleoside, and phosphate (see Table S1). For our experimental conditions, i.e., reaction times of 24 h at pH 9.0 and 40 °C, we ensured that the enzyme remained active and deoxyribose-1-phosphate did not degradate (see Figure S2).

To describe the recorded data, a differential–dynamical model was set up. This model allows for the simulation of the concentrations of substrates, products, and enzyme forms over an arbitrarily long time-course. The enzyme reaction can reach a dynamic equilibrium and assumes equal contribution of both substrates to reaction rates and level of equilibrium, as dictated by the underlying law of mass action. We conducted local optimization of the parameters $k = (k_1, k_{-1}, k_2, k_{-2})^T$ (given in unitless numbers for simplicity and in transposed vector form for brevity) to find a parameter set which described the data well. The parameter set we found to perform best on our experimental data is $k = (0.42, 0.17, 0.31, 7.6)^T$. The explicit form ($k_1 = 0.42$ (mM)$^{-2}$ min^{-1}; $k_{-1} = 0.17$ min^{-1}; $k_2 = 0.31$ min^{-1}; $k_{-2} = 7.6$ (mM)$^{-2}$ min^{-1}) will be omitted from here on for reasons of brevity.

In the tested range of enzyme and substrate concentrations, we found an excellent agreement between experimental data and our model with this parameter set (Figure 3). The predictions of our models were consistent and evenly distributed around the experimental data points over the whole time course of 24 h (Figures S3 and S4). We could not detect any particular trend of prediction errors towards phosphate, deoxythymidine, or enzyme concentrations. Thus, we conclude that our model is well balanced in the range of experimental conditions described here.

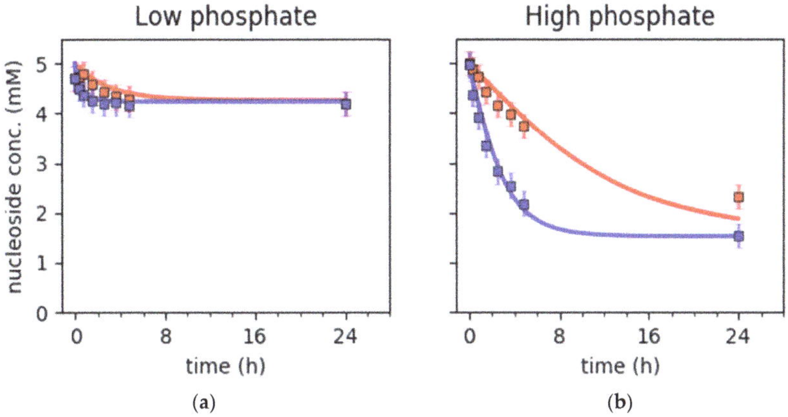

Figure 3. Exemplary fits for experimental data at low and high phosphate concentrations. (**a**) Experimental data and model predictions for conditions with low phosphate-to-deoxythymidine ratio (2 mM : 5 mM), and varying enzyme concentrations (red: High enzyme concentration, Experiment #13; blue: Low enzyme concentration, Experiment #12). Though the speed of reaction differs in the beginning, both reactions reach the same equilibrium during the time course of the experiment. Error bars represent 95% confidence intervals for the experimentally determined concentrations; (**b**) experimental data and model predictions for conditions with high phosphate-to-deoxythymidine ratio (80 mM : 5 mM), and varying enzyme concentrations (red: High enzyme concentration, Experiment #21; blue: low enzyme concentration, Experiment #20). The two conditions differ in their speed and low enzyme concentration is not sufficient to reach equilibrium. Error bars represent 95% confidence intervals for the experimentally determined concentrations. See Table S1 for experimental condition numbers as given in this figure legend ("Experiment #").

3.3. Multiple Parameter Sets Can Be Used for the Description of the Phosphorolysis Reaction

We performed global optimizations with basin-hop and differential evolution algorithms, as well as large-scale local optimizations from widely distributed initial parameter set guesses to find the best global parameter set. We found multiple parameter sets to describe the experimental dataset with almost similar accuracy. Except for k_2, which is almost constant, some alternative parameter sets, e.g., $k^* = (0.18, 0.12, 0.35, 5.5)^T$ or $k^{**} = (1.4, 0.94, 0.28, 4.6)^T$, differ drastically from the optimal parameter set $k = (0.42, 0.17, 0.31, 7.6)^T$. However, the cost functions are insignificantly different, with $Z(k) = 2.9 \times 10^3$, $Z(k^*) = 3.0 \times 10^3$, and $Z(k^{**}) = 3.2 \times 10^3$ (all values in (mM)^2). In practice, this can be attributed to the lacking difference in goodness-of-fit for comparison of simulations for k and the alternative parameter sets, as visualized in Figure 4.

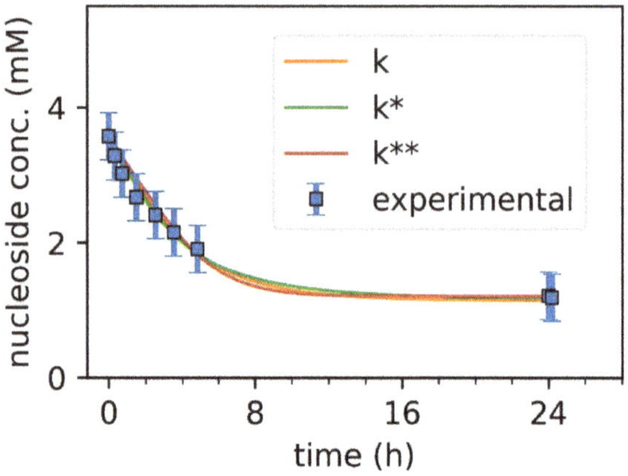

Figure 4. Non-identifiability of parameter sets from given experimental data. Experimental data (blue squares) and simulation of parameter sets k, k*, and k**, for a given experimental condition. The modelled results from different parameter sets are almost indiscernible, and, therefore, no decision can be taken on which parameter set is correct.

As the forward reaction is described reasonably well with multiple parameter sets, we cannot decide for any parameter set from our experimental data. This is caused by low sensitivities of the uncertain parameters in regard to our experimental data. As the scope of this work is fulfilled by a model for the forward reaction only, and all parameter sets describe the forward reaction reasonably well, we chose to communicate the parameter set k with lowest value of cost function Z.

3.4. *The Value of the Thermodynamic Equilibrium Constant Is Constant across Methods of Determination*

Finally, we investigated the behavior of the thermodynamic equilibrium constant across all experimental conditions. The thermodynamic equilibrium is approached when there is no observable change in the concentration of the enzyme complex, [EC]:

$$\frac{d[\mathrm{EC}]}{dt} = 0\,. \tag{11}$$

For our model, this yields two forms to express the equilibrium constant: Either (1) by considering the concentrations of substrates and products at equilibrium:

$$\mathrm{K_{eq}} = \frac{\mathrm{B_{eq}} \times \mathrm{S1P_{eq}}}{\mathrm{N_{eq}} \times \mathrm{P_{eq}}}\,, \tag{12}$$

or (2) by considering the parameter values:

$$\mathrm{K_{eq}} = \frac{\mathrm{k_1} \times \mathrm{k_2}}{\mathrm{k_{-1}} \times \mathrm{k_{-2}}}\,. \tag{13}$$

Estimating the equilibrium constant from the values found in the parameter estimation, one obtains $K_{eq} = 0.10$. The value of the equilibrium constant is approximately the same for alternative parameter sets, e.g., k* and k**, emphasizing the principal agreement between multiple parameter sets with the given experimental data.

Similarly, it is possible to derive the equilibrium constant from the equilibrium concentrations of products and substrates, giving a median value of $K_{eq} = 0.10$, similar to the value calculated from kinetic parameters (Figure 5).

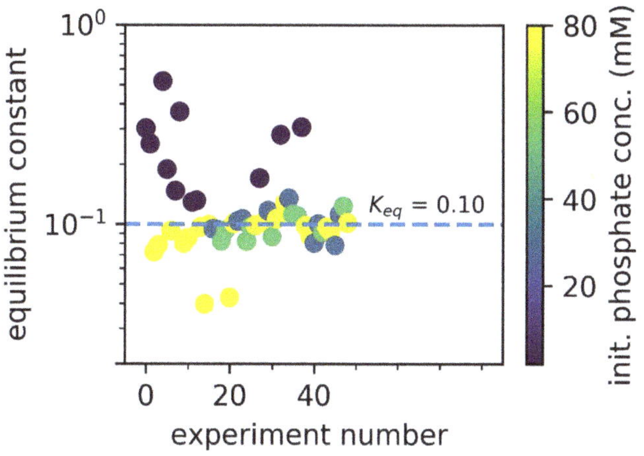

Figure 5. Equilibrium constant determined at different phosphate concentrations. Equilibrium constants of the 48 experiments under varying conditions (see Table S1) as determined from the deoxythymidine/thymine ratios of the 24 h data point as described in Materials and Methods. For concentrations of >2mM phosphate, the experimentally determined values from the 24 h data points are evenly distributed around the value calculated from the parameter estimation (K_{eq} = 0.10). For experiments with lower phosphate concentrations (dark purple circles), the equilibrium constant is significantly off the calculated value from the parameter sets. The median of all experiments is equal to the value calculated from the parameter estimation. Colored circles: K_{eq} calculated from the 24 h data points, purple to yellow: Increasing initial phosphate concentration; blue dashed line: K_{eq} = 0.10, as calculated from the parameter estimation.

4. Discussion

To the best of our knowledge, this study presents the first ODE model of an enzymatic two-substrate two-product process. For biotechnological production processes, it is desired to reach equilibrium state conditions to maximize the product yield. For the description of such processes, ODE models are required. In this contribution, we developed such a differential–dynamical model, which places a process perspective onto the enzymatic nucleoside phosphorylation reaction, and which is, regardless of its simplicity, in excellent agreement with our experimental data.

4.1. Model Structure

Contrary to Cleland's interpretation of multi-substrate/multi-product enzyme reactions [39–41], which considers multiple enzyme complex intermediates, we modeled the production process as consecutive law of mass actions, and only included one enzyme complex intermediate. Further, we explicitly decided to simplify a probable ordered binding mechanism [10] towards a three-particle collision. In our eyes, these simplifications are justified by the excellent agreement between the experimental data and our model (Figure 3 and Figure S3).

Further elegance of our model is found in its pluggability of equations, which allows for the easy introduction or decommissioning of individual reaction steps. Further, it does not need to rely on steady-state assumptions, although it is easy to integrate these. Finally, it is easier to provide explicit and precise description of, e.g., inhibitory actions into mechanistic models.

To date, our model does not include terms for the decay of enzyme activity or the degradation of any chemical species. We base these decisions on reports of the exceptional stability in alkaline conditions of deoxyribose-1-phosphate [34] and ribose-1-phosphate [42], as well as on the report of stable enzyme activity over days for thermophilic pyrimidine- [20] and purine-nucleoside phosphorylases [43] at even higher temperatures than those used in this study.

4.2. Plausibility of Our Results

To check the correctness of our results, we compared the equilibrium constant (1) from literature and (2) derived from our parameter sets or (3) determined by the equilibrium concentrations of the latest data points. As shown in the Results section, (2) and (3) accord with each other. The values for the equilibrium constant of the Py-NPase catalyzed reaction can be approximated by considering examples from literature [9,10,44,45], being in the similar order of magnitude for related reactions but differing in temperature, buffers, and exact specifications of base and sugar moiety. For a reaction with similar substrates at not too distant experimental conditions, the equilibrium constant was found to be K_{eq} = 0.102 at 37 °C and pH 7.4 [7]. This equals the equilibrium constant determined in this work.

Searching for experimental conditions that could discriminate between multiple parameter sets, we found major differences between the parameter sets to be only visible in kinetic study of the backward reaction. Exemplarily, parameter set k^{**} would show significantly faster conversion of deoxyribose-1-phosphate than parameter set k, as k_{-1} is significantly larger. The parameters k_{-1} and k_2 can be understood to correlate with the k_{cat} values of the phosphorolysis and synthesis reaction, respectively. Previous work [9] included the progress curve of one phosphorolysis and the corresponding synthesis reaction, and the initial reaction rates can be estimated from the graph given there, being of approximately similar speed. This argument favors k over k^{**}, but for k and k^* the situation is less clear. Further research needs to be conducted to resolve this ambiguity.

4.3. Limitations and Domain of Validity of the Model

The stability of deoxyribose-1-phosphate and enzyme activity over the assay duration is key to correct conclusions from the experimental data. In addition to reports from literature on pentose-1-phosphate and enzyme stability [20,34,42,43], we performed a control experiment via a coupled read-out, providing evidence for fulfilling these preconditions (Figure S2).

This also clearly points out the domain of validity of the model: In its current form, it applies only in the direction of phosphorolytic cleavage at pH 9.0 and 40 °C for time frames up to 24 h. Outside of this region, one should act on the assumption that corrections will be necessary not only for the temperature- and pH-dependence of the kinetic constants but also in the model structure regarding enzyme inactivation and reactant degradation, especially of deoxyribose-1-phosphate (Figure S3).

4.4. Application of Our Results to Production Processes

In the perspective of process control, our model has the potential to describe reactions in a time-resolved fashion, integrating knowledge which was previously not put into equations. The literature is rich in references of successful production processes with nucleoside phosphorylases, but these are typically focused on transglycosylations [35,46–48]. These processes are coupling two nucleoside-phosphorylase reactions, using pentose-1-phosphate as an intermediate in situ; however, for these processes, a prediction of time-resolved process performance was usually not undertaken.

A major advantage of a dynamic model is the ability to optimize processes before or during the run time. Exemplarily, one might want to minimize the amount of consumed enzyme for a batch-process with fixed run time. Our model allows for the calculation of the final yield and required enzyme amount for a fixed run-time, given constraints like, e.g., the solubility of substrate or limiting excess of phosphate (which, for the synthesis of pentose-1-phosphates, is typically used in 1- to 2.5-fold excess to ease down-stream processing). Similarly, one can calculate the run-time required to reach, e.g., 90% of equilibrium, given an amount of enzyme, substrate, and phosphate. These predictions are the basis of a cost-efficient production.

5. Conclusions and Outlook

The determination of the deoxythymidine/thymine ratios with UV/Vis spectroscopy is a fast and cost-effective method for assaying Py-NPase reactions. With this method in hand, we were able to set-up a model capable of describing the time course of Py-NPase reactions for the biotechnological production of deoxyribose-1-phosphate under diverse experimental conditions.

From this, we strive for the predictability of multi-step enzymatic reactions to produce nucleic acid derivatives. Our results pave the way for a significant improvement of production processes towards the synthesis of pharmaceutically interesting nucleosides.

Whenever available, time-resolved information on reaction progress can be used to parametrize the presented model structure. We believe that dynamic modelling will enable efficient process control and reaction engineering, especially when fully parametrized differential–dynamical models for nucleoside phosphorylation reactions are shared within the community.

Especially in multi-enzyme reactions, it will be necessary to integrate terms for undesired reactions, e.g., for product degradation or enzyme inactivation. Our model structure allows for an easy integration of additional terms ("coupling of models"). This would be much less feasible for traditional representations of enzyme kinetics, e.g., in Michaelis–Menten or Cleland notation.

After all, more studies on equilibrium constants and the relationship of kinetic rates at varying experimental conditions, e.g., temperatures or pH values, will be necessary to elucidate the mechanisms of this enzymatically catalyzed reaction further. Dynamic experiments, i.e., varying, for example, the temperature or concentration of reactants, can be next steps for the evaluation and refinement of our results.

Supplementary Materials: The following are available online at http://www.mdpi.com/2227-9717/7/6/380/s1 and https://doi.org/10.5281/zenodo.3243519, Figure S1: Further mixtures of deoxythymidine/thymine, and "predicted vs actual" plot, Table S1: Experimental conditions in this study, Figure S2: Degradation progress of deoxyribose-1-phosphate at elevated temperatures, Figure S3: Fits of all experiments, Figure S4: Comparison of inter-day controls.

Author Contributions: Conceptualization, R.T.G., M.N.C.B., A.W. and P.N.; Data curation, R.T.G., N.K. and F.K.; Formal analysis, R.T.G. and N.K.; Funding acquisition, R.T.G., M.N.C.B. and P.N.; Investigation, R.T.G., N.K. and F.K.; Methodology, R.T.G., N.K. and M.N.C.B.; Project administration, R.T.G.; Resources, R.T.G., M.N.C.B., A.W. and P.N.; Software, R.T.G. and N.K.; Supervision, R.T.G., M.N.C.B., P.N. and M.G.; Validation, F.K.; Visualization, R.T.G., F.K. and M.G.; Writing—original draft, R.T.G., F.K. and M.G.; Writing—review & editing, R.T.G., N.K., F.K., M.N.C.B., A.W., P.N. and M.G.

Funding: RTG was supported by the Berlin International Graduate School for Natural Sciences and Engineering (BIG-NSE) and the Einstein Center for Catalysis (EC^2). This research was funded by the Deutsche Forschungsgemeinschaft (DFG, German Research Foundation) under Germany's Excellence Strategy—EXC 2008/1 (UniSysCat)—390540038. We acknowledge support by the German Research Foundation and the Open Access Publication Fund of TU Berlin.

Acknowledgments: We thank Athel Cornish-Bowden and María Luz Cárdenas for inspiring discussions and valuable feedback on this manuscript. The authors are grateful to Sarah Kamel (TU Berlin and American University of Cairo) for providing Py-NPase, and to Sebastian Hans (TU Berlin) for sharing computational resources. We thank Mathis Gruber (DexLeChem GmbH) for fruitful discussion in the initial phase of this project.

Conflicts of Interest: AW is CEO of the biotech company BioNukleo GmbH. PN is member of the advisory board of BioNukleo GmbH. The funders had no role in the design of the study; in the collection, analyses, or interpretation of data; in the writing of the manuscript, or in the decision to publish the results.

References

1. Kamel, S.; Yehia, H.; Neubauer, P.; Wagner, A. Enzymatic synthesis of nucleoside analogues by nucleoside phosphorylases. In *Enzymatic and Chemical Synthesis of Nucleic Acid Derivatives*; Fernández–Lucas, J., Ed.; Wiley–VCH Verlag GmbH & Co. KGaA: Weinheim, Germany, 2018; pp. 1–28.
2. Lapponi, M.J.; Rivero, C.W.; Zinni, M.A.; Britos, C.N.; Trelles, J.A. New developments in nucleoside analogues biosynthesis: A review. *J. Mol. Catal. B Enzym.* **2016**, *133*, 218–233. [CrossRef]
3. Mikhailopulo, I.A.; Miroshnikov, A.I. New trends in nucleoside biotechnology. *Acta Nat.* **2010**, *2*, 36–59.
4. Del Arco, J.; Fernández–Lucas, J. Purine and pyrimidine salvage pathway in thermophiles: A valuable source of biocatalysts for the industrial production of nucleic acid derivatives. *Appl. Microbiol. Biotechnol.* **2018**, *102*, 7805–7820. [CrossRef] [PubMed]

5. Yehia, H.; Kamel, S.; Paulick, K.; Wagner, A.; Neubauer, P. Substrate spectra of nucleoside phosphorylases and their potential in the production of pharmaceutically active compounds. *Curr. Pharm. Des.* **2017**, *23*, 6913–6935. [CrossRef]
6. Cornish-Bowden, A. *Fundamentals of Enzyme Kinetics*, 4th ed.; Wiley: Weinheim, Germany, 2013; ISBN 9783527665488.
7. Bose, R.; Yamada, E.W. Uridine phosphorylase, molecular properties, and mechanism of catalysis. *Biochemistry* **1974**, *13*, 2051–2056. [CrossRef]
8. Hori, N.; Uehara, K.; Mikami, Y. Effects of Xanthine Oxidase on Synthesis of 5–Methyluridine by the Ribosyl Transfer Reaction. *Agric. Biol. Chem.* **1991**, *55*, 1071–1074. [CrossRef]
9. Friedkin, M. Desoxyribose–1–phosphate: II. The isolation of crystalline desoxyribose–1–phosphate. *J. Biol. Chem.* **1950**, *184*, 449–459.
10. Porter, D.J. Purine nucleoside phosphorylase: Kinetic mechanism of the enzyme from calf spleen. *J. Biol. Chem.* **1992**, *267*, 7342–7351.
11. Jeoh, T.; Cardona, M.J.; Karuna, N.; Mudinoor, A.R.; Nill, J. Mechanistic kinetic models of enzymatic cellulose hydrolysis: A review. *Biotechnol. Bioeng.* **2017**, *114*, 1369–1385. [CrossRef] [PubMed]
12. Shin, J.S.; Kim, B.G. Kinetic modeling of ω–transamination for enzymatic kinetic resolution of α–methylbenzylamine. *Biotechnol. Bioeng.* **1998**, *60*, 534–540. [CrossRef]
13. Van Hecke, W.; Bhagwat, A.; Ludwig, R.; Dewulf, J.; Haltrich, D.; Van Langenhove, H. Kinetic modeling of a bi–enzymatic system for efficient conversion of lactose to lactobionic acid. *Biotechnol. Bioeng.* **2009**, *102*, 1475–1482. [CrossRef] [PubMed]
14. Zavrel, M.; Schmidt, T.; Michalik, C.; Ansorge–Schumacher, M.; Marquardt, W.; Büchs, J.; Spiess, A.C. Mechanistic kinetic model for symmetric carboligations using benzaldehyde lyase. *Biotechnol. Bioeng.* **2008**, *101*, 27–38. [CrossRef] [PubMed]
15. Levenspiel, O. Chemical Reaction Engineering. *Ind. Eng. Chem. Res.* **1999**, *38*, 4140–4143. [CrossRef]
16. Franceschini, G.; Macchietto, S. Model–based design of experiments for parameter precision: State of the art. *Chem. Eng. Sci.* **2008**, *63*, 4846–4872. [CrossRef]
17. Asprey, S.P.; Macchietto, S. Designing robust optimal dynamic experiments. *J. Process Control* **2002**, *12*, 545–556. [CrossRef]
18. Wilms, T.; Rischawy, D.F.; Barz, T.; Esche, E.; Repke, J.U.; Wagner, A.; Neubauer, P.; Cruz Bournazou, M.N. Dynamic optimization of the PyNP/PNP phosphorolytic enzymatic process using MOSAICmodeling. *Chem. Ing. Tech.* **2017**, *89*, 1523–1533. [CrossRef]
19. Krausch, N.; Barz, T.; Sawatzki, A.; Gruber, M.; Kamel, S.; Neubauer, P.; Cruz Bournazou, M.N. Monte Carlo Simulations for the Analysis of Non–linear Parameter Confidence Intervals in Optimal Experimental Design. *Front. Bioeng. Biotechnol.* **2019**, *7*, 4747. [CrossRef]
20. Szeker, K.; Zhou, X.; Schwab, T.; Casanueva, A.; Cowan, D.; Mikhailopulo, I.A.; Neubauer, P. Comparative investigations on thermostable pyrimidine nucleoside phosphorylases from Geobacillus thermoglucosidasius and Thermus thermophilus. *J. Mol. Catal. B Enzym.* **2012**, *84*, 27–34. [CrossRef]
21. Yamada, E.W. Pyrimidine nucleoside phosphorylases of rat liver. Separation by ion exchange chromatography and studies of the effect of cytidine or uridine administration. *J. Biol. Chem.* **1968**, *243*, 1649–1655.
22. Shibata, S.; Furukawa, M.; Goto, K. Dual–wavelength spectrophotometry. *Anal. Chim. Acta* **1969**, *46*, 271–279. [CrossRef]
23. Mohamed, E.H.; Lotfy, H.M.; Hegazy, M.A.; Mowaka, S. Different applications of isosbestic points, normalized spectra and dual wavelength as powerful tools for resolution of multicomponent mixtures with severely overlapping spectra. *Chem. Cent. J.* **2017**, *11*, 43. [CrossRef] [PubMed]
24. Python Core Team. The Python Language Reference, version 2.7. Available online: https://docs.python.org/2.7/reference/index.html (accessed on 11 June 2019).
25. Python Core Team. The Python Language Reference, version 3.6. Available online: https://docs.python.org/3.6/reference/index.html (accessed on 11 June 2019).
26. Krausch, N.; Giessmann, R.T. Collection of UV/Vis Spectra for Monitoring of Reaction Progress with Varying Reactant Concentrations, version 1.0.0. [Data Set]. Available online: https://doi.org/10.5281/zenodo.3243352 (accessed on 11 June 2019).
27. Giessmann, R.T.; Krausch, N. Data Toolbox, version 0.11.1. [Software]. Available online: https://doi.org/10.5281/zenodo.3243361 (accessed on 11 June 2019).

28. Giessmann, R.T. Mbdoe–Python, version 0.1.1. [Software]. Available online: https://doi.org/10.5281/zenodo.3243374 (accessed on 11 June 2019).
29. Giessmann, R.T. Gitflower, version 0.3.0. [Software]. Available online: https://doi.org/10.5281/zenodo.3243368 (accessed on 11 June 2019).
30. Meurer, A.; Smith, C.P.; Paprocki, M.; Čertík, O.; Kirpichev, S.B.; Rocklin, M.; Kumar, A.; Ivanov, S.; Moore, J.K.; Singh, S.; et al. SymPy: Symbolic computing in Python. *PeerJ Comput. Sci.* **2017**, *3*, e103. [CrossRef]
31. Verdier, O.; Tenfjord, R. Python–Sundials. Available online: https://github.com/olivierverdier/python-sundials (accessed on 24 May 2019).
32. Hindmarsh, A.C.; Brown, P.N.; Grant, K.E.; Lee, S.L.; Serban, R.; Shumaker, D.E.; Woodward, C.S. SUNDIALS. *ACM Trans. Math. Softw.* **2005**, *31*, 363–396. [CrossRef]
33. Newville, M.; Stensitzki, T.; Allen, D.B.; Ingargiola, A. LMFIT, version 0.9.12. [Software]. Available online: https://doi.org/10.5281/zenodo.11813 (accessed on 11 June 2019).
34. Kamel, S.; Weiß, M.; Klare, H.F.T.; Mikhailopulo, I.A.; Neubauer, P.; Wagner, A. Chemo–enzymatic synthesis of α–D–pentofuranose–1–phosphates using thermostable pyrimidine nucleoside phosphorylases. *Mol. Catal.* **2018**, *458*, 52–59. [CrossRef]
35. Ubiali, D.; Rocchietti, S.; Scaramozzino, F.; Terreni, M.; Albertini, A.M.; Fernández–Lafuente, R.; Guisán, J.M.; Pregnolato, M. Synthesis of 2′–deoxynucleosides by transglycosylation with new immobilized and stabilized uridine phosphorylase and purine nucleoside phosphorylase. *Adv. Synth. Catal.* **2004**, *346*, 1361–1366. [CrossRef]
36. Stimson, M.M.; Reuter, M.A. The Effect of pH on the spectra of thymine and thymine desoxyriboside. *J. Am. Chem. Soc.* **1945**, *67*, 847–848. [CrossRef]
37. Shugar, D.; Fox, J.J. Spectrophotometric studies op nucleic acid derivatives and related compounds as a function of pH. *Biochim. Biophys. Acta* **1952**, *9*, 199–218. [CrossRef]
38. Fox, J.J.; Shugar, D. Spectrophotometric studies on nucleic acid derivatives and related compounds as a function of pH: II. Natural and synthetic pyrimidine nucleosides. *Biochim. Biophys. Acta* **1952**, *9*, 369–384. [CrossRef]
39. Cleland, W.W. The kinetics of enzyme–catalyzed reactions with two or more substrates or products: I. Nomenclature and rate equations. *Biochim. Biophys. Acta* **1963**, *67*, 104–137. [CrossRef]
40. Cleland, W.W. The kinetics of enzyme–catalyzed reactions with two or more substrates or products: II. Inhibition: Nomenclature and theory. *Biochim. Biophys. Acta* **1963**, *67*, 173–187. [CrossRef]
41. Cleland, W.W. The kinetics of enzyme-catalyzed reactions with two or more substrates or products: III. Prediction of initial velocity and inhibition patterns by inspection. *Biochim. Biophys. Acta* **1963**, *67*, 188–196. [CrossRef]
42. Halmann, M.; Sanchez, R.A.; Orgel, L.E. Phosphorylation of D–ribose in aqueous solution. *J. Org. Chem.* **1969**, *34*, 3702–3703. [CrossRef]
43. Zhou, X.; Szeker, K.; Janocha, B.; Böhme, T.; Albrecht, D.; Mikhailopulo, I.A.; Neubauer, P. Recombinant purine nucleoside phosphorylases from thermophiles: Preparation, properties and activity towards purine and pyrimidine nucleosides. *FEBS J.* **2013**, *280*, 1475–1490. [CrossRef] [PubMed]
44. Kalckar, H.M. The enzymatic synthesis of purine ribosides. *J. Biol. Chem.* **1947**, *167*, 477–486. [PubMed]
45. Yamada, E.W. Uridine phosphorylase from rat liver. *Methods Enzymol.* **1978**, *51*, 423–431. [CrossRef]
46. Zhou, X.; Szeker, K.; Jiao, L.Y.; Oestreich, M.; Mikhailopulo, I.A.; Neubauer, P. Synthesis of 2,6–dihalogenated purine nucleosides by thermostable nucleoside phosphorylases. *Adv. Synth. Catal.* **2015**, *357*, 1237–1244. [CrossRef]

47. Zhou, X.; Yan, W.; Zhang, C.; Yang, Z.; Neubauer, P.; Mikhailopulo, I.A.; Huang, Z. Biocatalytic synthesis of seleno, thio and chloro–nucleobase modified nucleosides by thermostable nucleoside phosphorylases. *Catal. Commun.* **2019**, *121*, 32–37. [CrossRef]
48. Cattaneo, G.; Rabuffetti, M.; Speranza, G.; Kupfer, T.; Peters, B.; Massolini, G.; Ubiali, D.; Calleri, E. Synthesis of adenine nucleosides by transglycosylation using two sequential nucleoside phosphorylase–based bioreactors with on–line reaction monitoring by using HPLC. *ChemCatChem* **2017**, *9*, 4614–4620. [CrossRef]

processes

MDPI

Article

Dynamic Modelling and Optimisation of the Batch Enzymatic Synthesis of Amoxicillin

Andrew B. Cuthbertson, Alistair D. Rodman, Samir Diab and Dimitrios I. Gerogiorgis *

Institute for Materials and Processes (IMP), School of Engineering, University of Edinburgh, The Kings Buildings, Edinburgh, Scotland EH9 3FB, UK; s1533497@sms.ed.ac.uk (A.B.C.); A.Rodman@ed.ac.uk (A.D.R.); S.Diab@ed.ac.uk (S.D.)

* Correspondence: D.Gerogiorgis@ed.ac.uk; Tel.: +44-131-651-7072

Received: 7 May 2019; Accepted: 24 May 2019; Published: 28 May 2019

Abstract: Amoxicillin belongs to the β-lactam family of antibiotics, a class of highly consumed pharmaceutical products used for the treatment of respiratory and urinary tract infections, and is listed as a World Health Organisation (WHO) "Essential Medicine". The demonstrated batch enzymatic synthesis of amoxicillin is composed of a desired synthesis and two undesired hydrolysis reactions of the main substrate (6-aminopenicillanic acid (6-APA)) and amoxicillin. Dynamic simulation and optimisation can be used to establish optimal control policies to attain target product specification objectives for bioprocesses. This work performed dynamic modelling, simulation and optimisation of the batch enzymatic synthesis of amoxicillin. First, kinetic parameter regression at different operating temperatures was performed, followed by Arrhenius parameter estimation to allow for non-isothermal modelling of the reaction network. Dynamic simulations were implemented to understand the behaviour of the design space, followed by the formulation and solution of a dynamic non-isothermal optimisation problem subject to various product specification constraints. Optimal reactor temperature (control) and species concentration (state) trajectories are presented for batch enzymatic amoxicillin synthesis.

Keywords: Amoxicillin; enzymatic synthesis; non-isothermal modelling; parameter estimation; dynamic optimisation

1. Introduction

Antibiotics are societally essential pharmaceutical products, making previously untreatable illnesses such as pneumonia and tuberculosis curable, thus revolutionising modern medicine [1]. Access to essential medicines in low- to middle-income countries and antibiotic shortages in developed countries remain an important issue [2,3]. The complex molecular structures of antibiotics imply expensive, multistep and materially intensive syntheses required to make such molecules [4]. Designing efficient and cost effective antibiotic manufacturing routes is imperative [5].

The family of β-lactam antibiotics includes some of the most important pharmaceutical products, with cephalosporins and semisynthetic penicillins corresponding to around 65% of the global production of antibiotics [6]. Figure 1 illustrates the leading antibiotics by share of infection treatments in the U.K. in 2016, with five of the leading nine antibiotics belonging to the β-lactam family. Table 1 lists certain β-lactam antibiotics with their applications, average sales volumes [7,8] and unit prices, taken from the National Chemical Database Service. The production of high-sale-volume β-lactam antibiotics is typically implemented via enzymatic methods [9]. Amoxicillin is one β-lactam antibiotic listed as a World Health Organisation (WHO) "Essential Medicine": It is applicable to a variety of ailments [10], including respiratory and urinary tract infections (UTIs) and is the top antibiotic in terms of dosage worldwide [11]. The demonstrated enzymatic synthesis of amoxicillin paves the way for the elucidation of optimal design and operating parameters via modelling and optimisation [12,13].

Dynamic modelling and optimisation of batch processes have been substantially implemented in the literature for a variety of bioprocesses, including fermentations [14,15], antibiotic synthesis [16,17] and many other applications, illustrating the utility and versatility of such methods in bioprocess design. Enzymatic synthesis, crystallisation and reactive crystallisation studies for β-lactam antibiotics have been implemented in the literature [16–19]. A demonstrated batch enzymatic synthesis of amoxicillin paves the way for theoretical studies [20]. Dynamic modelling and optimisation of the batch enzymatic synthesis of amoxicillin has yet to be implemented. Moreover, introducing temperature dependence into the model may allow for the optimisation of batch reactor temperature profiles to meet a specific production objective, which, to the best of our knowledge, has not been previously implemented.

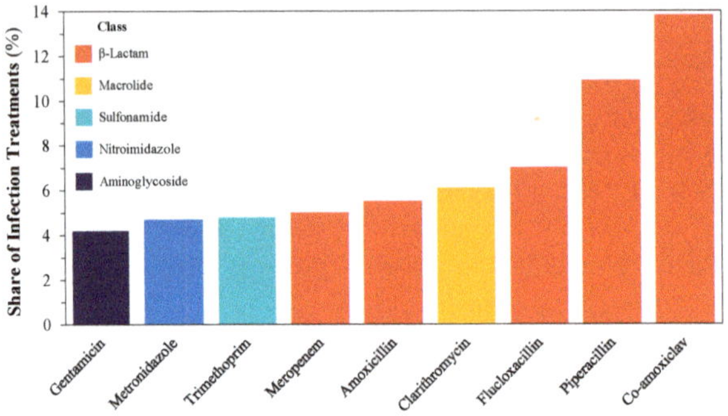

Figure 1. Leading antibiotics used in the U.K. in 2016 [21].

The main objectives and novelty of this work are as follows:

1. To introduce temperature dependency into the published kinetic model of batch enzymatic amoxicillin synthesis;
2. To understand the attainable performances and inherent trade-offs (via isothermal operation) of varying batch times and operating temperatures;
3. To optimise dynamic temperature profiles toward optimal process performance for varying product quality constraints.

First, the enzymatic synthesis pathway is presented with the kinetic model. The temperature dependence of kinetic parameters is introduced through the regression of Arrhenius constants from published experimental data. An isothermal simulation is then implemented for design space investigation. A constrained dynamic optimisation problem is then described, which compares the resulting temperature (control) profiles required to meet certain product specification objectives pertinent to the specific process.

2. Dynamic Modelling and Optimisation

2.1. *Amoxicillin Synthesis Pathway and Kinetic Model*

The kinetically controlled reaction pathway for the synthesis of amoxicillin catalysed by penicillin G acylase (PGA, denoted as E) is shown in Figure 2. A classical mechanistic approach to the complete mechanism for this system has been shown to likely lead to an intractable problem: Thus, a simplified semi-empirical kinetic model is used. Amoxicillin (AMOX) is synthesised from the reaction of p-hydroxyphenyl glycine methyl ester (PHPGME) and 6-aminopenicillanic acid (6-APA). There are two side reactions: PHPGME hydrolysis to p-hydroxyphenyl glycine (PHPG) and amoxicillin hydrolysis to 6-APA and PHPG. Pertinent species properties are listed in Table 2. The enzyme concentration, C_E, equals that considered in the literature [20].

Table 1. Different β-lactam antibiotic subclasses, applications, properties and economic data.

Subclass	Antibiotic	Spectrum	Generation	Application	CAS #	MW	Sales [7,8]	Price
						(g mol^{-1})	(tonnes)	(10^3 GBP kg^{-1})
Penicillin	Penicillin V	Narrow	1st	Laryngitis, bronchitis, pneumonia, skin infections	87-08-1	350.39	126.29	82.584
	Oxacillin	Narrow	2nd	Staphylococci infections	7240-38-2	401.44	2.87	293.334
	Nafcillin	Narrow	2nd	Staphylococci infections	985-16-0	414.48	8.48	27.206
	Dicloxacillin	Narrow	2nd	Bronchitis, pneumonia, staphylococci infections	3116-76-5	470.33	7.35	71.100
	Ampicillin	Broad	3rd	UTIs, pneumonia, gonorrhoea, meningitis, abdominal infections	69-53-4	349.41	42.35	11.358
	Amoxicillin	Broad	3rd	Tonsillitis, bronchitis, pneumonia, gonorrhoea, sinus infections, UTIs	26787-78-0	365.40	1,122.41	37.057
	Ticarcillin	Broad	4th	UTIs, bone + joint infections, stomach infections, skin infections	34787-01-4	384.43	2.92	46.487
	Piperacillin	Broad	4th	UTIs, bone + joint infections, stomach infections, skin infections	66258-76-2	517.56	140.51	60.854
Cephalosporin	Cephalexin	n/a	1st	UTIs, upper respiratory tract infections, ear infections, skin infections	15686-71-2	347.39	321.90	63.152
	Cefadroxil	n/a	1st	UTIs, staphylococci infections, skin infections	66592-87-8	363.39	11.75	83.638
	Cefazolin	n/a	1st	Respiratory tract infections, UTIs, skin infections,	25953-19-9	454.51	39.39	5.930
	Cefdinir	n/a	2nd	Bronchitis, pneumonia, skin infections, sinus infections	91832-40-5	395.42	41.77	59.457
	Cefaclor	n/a	2nd	UTIs, respiratory tract infections, sinus infections	53994-73-3	367.81	2.95	33.487
	Cefprozil	n/a	2nd	Ear infections, skin infections	92665-29-7	389.43	11.12	33.813
	Cefoxitin	n/a	2nd	UTIs, skin infections, pneumonia, bronchitis, tonsillitis, ear infections	35607-66-0	427.45	4.30	156.519
	Cefixime	n/a	3rd	Sinus infections, bronchitis, pneumonia	79350-37-1	453.45	1.73	59.919
	Cefotaxime	n/a	3rd	UTIs, pneumonia, abdominal infections, bone + joint infections	63527-52-6	455.47	2.57	251.725
	Ceftriaxone	n/a	3rd	Lower respiratory tract infections, UTIs, skin infections	104376-79-6	554.58	29.90	212.500
Monobactam	Aztreonam	n/a	n/a	Pneumonia	1-5-7	435.43	3.72	39.492

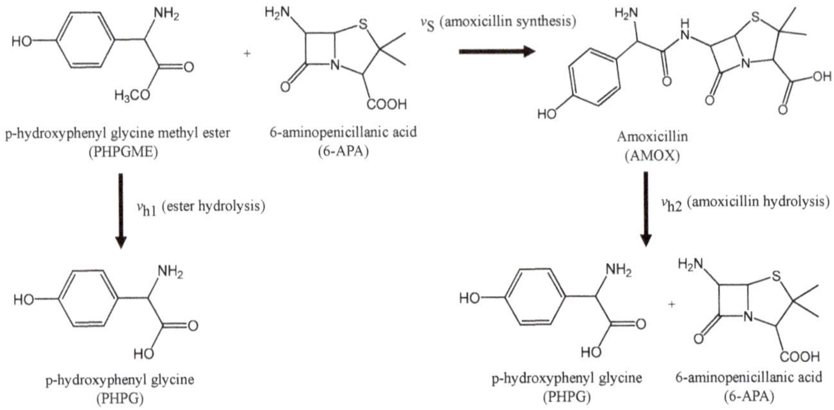

Figure 2. Simplified reaction pathway presented for the semiempirical model [20].

Equations (1)–(8) describe the kinetic model for the reaction pathways shown in Figure 2 [20]. Here, C = the species concentration at time t; v_i = the species rate of formation; v_{h1} = the rate of PHPGME hydrolysis; v_{h2} = the rate of AMOX hydrolysis; vS = the rate of AMOX synthesis; k = the species inhibition constant; k_{EN} = the 6-APA adsorption constant; k_{cat} = the reaction rate constant; K_M = the reaction empirical rate constant; and X_{max} is the maximum conversion ratio of the enzyme reagent complex into AMOX. Subscripts i and j denote species and reactions, respectively. The system of dynamic ODEs is solved simultaneously using the built-in MATLAB ODE solver ode15s. Equations (1)–(8) are

$$\frac{dC_i}{dt} = v_i \tag{1}$$

$$v_{AMOX} = v_S - v_{h2} \tag{2}$$

$$v_{6-APA} = v_{h2} - v_S \tag{3}$$

$$v_{PHPG} = v_{h1} + v_{h2} \tag{4}$$

$$v_{PHPGME} = \frac{k_{cat,2} C_E C_{PHPGME}}{K_{M1}\left(1 + \frac{C_{AMOX}}{k_{AMOX}} + \frac{C_{PHPG}}{k_{PHPG}}\right) + C_{PHPGME}} \tag{5}$$

$$v_{h1} = v_{PHPGME} - v_S \tag{6}$$

$$v_{h2} = \frac{k_{cat,1} C_E C_{AMOX}}{K_{M2}\left(1 + \frac{C_{PHPGME}}{k_{PHPGME}} + \frac{C_{6-APA}}{k_{6-APA}} + \frac{C_{PHPG}}{k_{PHPG}}\right) + C_{AMOX}} \tag{7}$$

$$v_S = \frac{k_{cat,2} C_E C_{PHPGME}}{K_{M1}\left(1 + \frac{C_{AMOX}}{k_{AMOX}} + \frac{C_{PHPG}}{k_{PHPG}}\right) + C_{AMOX}} \frac{C_E}{k_E + C_E} X_{max} \tag{8}$$

2.2. Kinetic Parameter Estimation

In this section, temperature dependency is introduced to the published kinetic model through the regression of Arrhenius parameters for certain parameters in Equations (1)–(8). The considered operating conditions of the batch reactor for enzymatic amoxicillin synthesis are as described in the experimental demonstration in the literature [22]. A 2:1 molar mixture of PHPGME and 6-APA (40:20 mM, respectively) in a 1-L batch volume was used for the batch enzymatic synthesis, with no amoxicillin or PHPG present at the start of the reaction, i.e., C_{PHPG} (t_0 = 0), C_{AMOX} (t_0 = 0) = 0.

Table 2. Properties of species in the reaction scheme for the batch enzymatic synthesis of amoxicillin.

Compound	Abbreviation	Type	CAS #	MW (g mol^{-1})
p-hydroxyphenyl glycine methyl ester	PHPGME	Feed	127369-30-6	180.18
6-aminopenicillanic acid	6-APA	Feed/Side product	551-16-6	216.26
amoxicillin	AMOX	Product	26787-78-0	365.40
p-hydroxyphenyl glycine methyl ester	PHPG	Side product	37784-25-1	167.16

The original dynamic model for the batch enzymatic synthesis of amoxicillin assumes isothermal operation, i.e., it does not account for the temperature dependence of kinetic parameters. Here, the kinetic parameter temperature dependence for amoxicillin is introduced from published temperature-dependent data: Values of $k_{cat,1}$ and $k_{cat,2}$ were regressed from published amoxicillin concentration data at 5, 25, and 35 °C [22], from which Arrhenius parameters could then be estimated. The remaining kinetic parameters were assumed to be temperature-independent, as a much wider kinetic dataset is required for further multiparametric regression. Values of $k_{cat,1}$ and $k_{cat,2}$ at different temperatures were regressed by minimising the residual error between the kinetic model (Equations (1)–(8)) and the experimental data using the bound constrained solver "fminsearchbnd" in MATLAB. Table 3 shows the regressed values of $k_{cat,1}$ and $k_{cat,2}$ at the given temperatures.

Table 3. Values of kinetic parameters under varying isothermal conditions.

	Parameter	$T = 5$ °C	$T = 25$ °C	$T = 35$ °C
T-dependent	$k_{cat,1}$ (IU g^{-1} min^{-1})	0.57	0.59	0.64
	$k_{cat,2}$ (IU g^{-1} min^{-1})	9.16	3.07	1.77
Fixed	K_{M1} (mM)	0.20	0.20	0.20
	K_{M2} (mM)	27.47	27.47	27.47
	X_{max} (–)	0.96	0.96	0.96
	k_E (mM)	16.03	16.03	16.03
	k_{PHPGME} (mM)	2672.04	2672.04	2672.04
	k_{AMOX} (mM)	4.59	4.59	4.59
	k_{PHPG} (mM)	4.51	4.51	4.51
	$k_{6\text{-}APA}$ (mM)	4550.28	4550.28	4550.28

With temperature-varying values for $k_{cat,1}$ and $k_{cat,2}$, the Arrhenius parameters, k_0 (the pre-exponential factor) and E_a (the energy barrier), were then regressed. An Arrhenius-type temperature dependence of $k_{cat,1}$ and $k_{cat,2}$ was assumed, according to Equation (9):

$$k_{cat}(T) = k_0 \exp\left(-\frac{E_a}{RT}\right) \tag{9}$$

where k_0 and E_a are the Arrhenius pre-exponential factor and energy barrier, respectively; R is the universal gas constant; and T is the reaction temperature. The fitting methodology for Arrhenius constant regression also used "fminsearchbnd" in MATLAB, as described previously. Figure 3 shows the lines of best fit for both Arrhenius plots, showing a good fit in both cases. The regressed Arrhenius parameter values (listed in Table 4) allowed for good replication of the experimental data (Figure 4). A corroboration of the regressed parameters with a wider dataset will further validate the values used in this work. The experimental data in the literature that were used to regress the kinetic parameters at different operating temperatures (T = 5, 25, 35 °C) provided no error values for the calculated concentrations (estimated via HPLC) or temperatures, and thus error bars are not shown [22]. An investigation of the effect of errors on regressed parameter values (isothermal $k_{cat,1}$ and $k_{cat,2}$ and non-isothermal Arrhenius parameters) could be implemented by perturbing their values and observing the effect on the optimisation results as a form of sensitivity analysis.

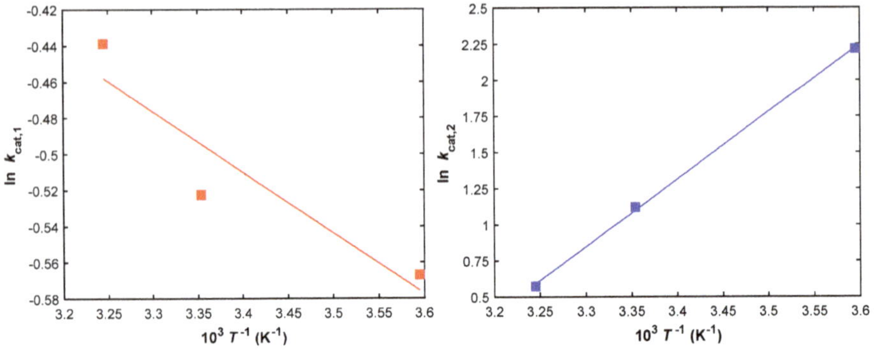

Figure 3. Arrhenius parameter regression from experimental data [22].

Table 4. Regressed Arrhenius parameters for $k_{cat,1}$ and $k_{cat,2}$.

Parameter	$k_{cat,1}$	$k_{cat,2}$
k_0 (IU g^{-1} min^{-1})	1.89	4.74×10^{-7}
E_a (J mol^{-1})	2.80×10^3	-3.88×10^4

2.3. *Dynamic Simulation and Optimisation*

2.3.1. Design Space Investigation and Simulation

Surface plots of the concentrations of key species, PHPGME, 6-APA, AMOX and PHPG as a function of isothermal reactor temperature and time were developed for design space investigation. Additionally, various performance indices were investigated and compared to elucidate attainable performances and inherent trade-offs. Selectivity, S, is the ratio of concentration of AMOX to PHPG (Equation (10)), and productivity, P, is the rate of production of AMOX (Equation (11)). Surface plots of these performance indices were also developed to understand the behaviour of the dynamic system.

$$S = \frac{C_{AMOX}}{C_{PHPG}} \tag{10}$$

$$P = \frac{C_{AMOX}^{max}}{t_{max}} \tag{11}$$

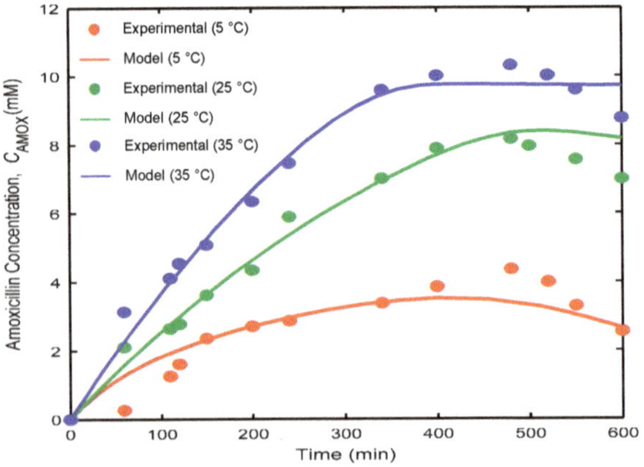

Figure 4. Concentration profiles of amoxicillin: experimental [22] vs. Arrhenius modelled results.

2.3.2. Dynamic Optimisation

The optimisation problem is posed such that different target amoxicillin product concentrations are met while consumption of the valuable feedstock 6-APA (bulk price = 35–40 USD kg^{-1}) is minimised. Minimising feedstock consumption in different cases can lead to possible process configurations where reagents and solvent are recycled following desired product (amoxicillin) crystallisation [23,24]. Although recycling options are not explicitly considered in this work, they have been demonstrated in the literature for β-lactam antibiotic synthesis with nanofiltration for by-product (PHPG) removal [25]. This can allow for processes with greater material efficiencies and overall sustainability. Economic analyses of specific process designs can further elucidate the benefits of such designs [5], but are beyond the scope of this work.

The problem considers system state variables, i.e., species concentration profiles ($C(t)$) that vary over time and are influenced by the reactor temperature control vector $T(t)$. The problem aims to simultaneously consider the minimisation of 6-APA consumption with the scope of potential recycling of this key substrate along with amoxicillin product maximisation. The bicriteria problem is defined by Equations (13) and (14). While J_1 and J_2 present competing objectives, the utility of the presented optimisation results is in exploring how to manipulate the optimal temperature control profile over the batch duration to yield the desired amount of amoxicillin while minimising feedstock consumption. Equations (12)–(14) are

$$\min_{T(t)} \begin{cases} J_1 \\ J_2 \end{cases} \tag{12}$$

$$J_1 = C_{6\text{-APA}}(t_0) - C_{6\text{-APA}}(t_f) \tag{13}$$

$$J_2 = -C_{AMOX}(t_f) \tag{14}$$

Numerous approaches can be used to modify a multiobjective problem for compatibility with single-objective solution methods. Commonly, a weighted sum objective is used to combine competing objectives into a single term, with weights defining the relative importance of each. However, the weights assigned to various process targets to produce a single objective function may be considered arbitrary in many cases, with decision-makers not necessarily able to quantify a priori the relative importance of competing objectives. Rather, we elected to consider an ε-constraint approach [26]. One of the objectives can be considered to be a constraint in the problem formulation, while the other is solved to optimality. This is repeated, increasing the value of the objective constraint by $\Delta\varepsilon$ and re-solving, repeating this process by incrementally increasing the constraint value across the entire span of permissible values for that particular objective. In this case, the amoxicillin product concentration was treated as a secondary objective and converted into a constraint. The single objective herein was to minimise the consumption of key feedstock 6-APA (formulated as the maximisation of final 6-APA concentration), subject to varying final amoxicillin concentrations (end-point constraints) and a final batch time, t_f = 500 min, which was observed as the maximum time beyond which amoxicillin degradation via hydrolysis to 6-APA and PHPG dominated (Figures 3 and 4). Equations (12)–(14) become Equations (15)–(18):

$$\max_{T(t)} J = C_{6-APA}(t_f) \tag{15}$$

s.t.

$$C_{AMOX}(t_f) \geq \varepsilon \tag{16}$$

$$t_f = 500 \text{ min} \tag{17}$$

$$5\ ^\circ\text{C} \leq T(t) \leq 35\ ^\circ\text{C} \tag{18}$$

Reactor temperature control trajectories were considered to be piecewise constant with N = 20 time discretisation elements. A variety of different initialisation profiles were tested to investigate the sensitivity of the problem to the initial guess of temperature trajectory, which consistently converged to the same optima presented in this work. Although global optima were not guaranteed, a variety of initialisation profiles resulting in the same optima and profiles implied that as close to global optima were attained as possible with the solution methods implemented. Temperature control profiles were initialised as isothermal at T = 295 K across all time discretisations. Lower and upper boundaries on temperature values were 5 and 35 °C, respectively (interior constraints, Equation (18)), corresponding to the temperature range used to regress kinetic parameter Arrhenius constants. Different end-point constraints on amoxicillin concentration (Equation (16)) were considered, as well as an unconstrained case ($\varepsilon = 0$). Each case of a considered amoxicillin concentration constraint was solved as a separate optimisation problem. The problem was solved using a direct, simultaneous method: Orthogonal collocation on finite elements [27,28] to approximate the state ($C(t)$) and control ($T(t)$) trajectories to convert the problem into a nonlinear programme (NLP), which was then solved using an interior point filter line search algorithm (IPOPT) [29]. Dynamic optimisation was implemented in MATLAB using DynOpt [30].

3. Results and Discussion

3.1. Dynamic Simulation

Concentration surface plots of PHPGME, 6-APA, AMOX and PHPG as a function of batch time and reactor operating temperature are shown in Figure 5. The colours in Figure 5 represent concentration data values on the z axis and are shown to aid in the interpretation of the surface plots. Species PHPGME and PHPG concentrations were strong functions of batch runtime but were largely unaffected by temperature, which was likely due to the model being fitted exclusively to AMOX concentration data [22]. Key substrate (6-APA and PHPGME) concentrations were lowest at longer batch times and higher temperatures, as these conditions favoured consumption toward amoxicillin and its hydrolysis to PHPG. Amoxicillin synthesis was favoured by longer batch times and higher reactor temperatures, while PHPG concentration was not affected significantly by reactor temperature. The concentration of AMOX reached a maximum at ~430 min, after which it began to decrease due to the decreasing ratios of vS to v_{h2} and v_{AMOX} to v_{PHPG} (as shown in Figure 6), both of which decreased significantly with batch time. The colours in Figure 6 represent the reaction rate ratios shown on the z axis and are shown to aid in the interpretation of the surface plots.

3.2. Non-Isothermal Simulation

Extensive simulations were performed as a preliminary investigation of attainable process performances for non-isothermal batch reactor operations. Here, the $T(t)$ domain was discretised on a coarse grid (discretisation level N = 9) to generate a finite set of profiles for exhaustive simulation. Possible temperature profiles were subject to the constraint that the temperature was allowed to change by a maximum of 10 °C per interval. The different simulation cases are presented in Figure 7, showing the maximum attainable amoxicillin concentrations versus the maximum batch time required (t_{MAX}), and also selectivity (Equation (10)) and productivity (Equation (11)). The banding observed was due to stepwise temperature profile simulations. Different cases are highlighted on each of these plots: (1) maximum selectivity, (2) maximum amoxicillin concentration and (3) a compromise between selectivity and productivity. Maximum final amoxicillin concentration was achieved by operating at the maximum temperature (35 °C), in agreement with previously observed results [22]. There are inherent trade-offs between different process performance metrics. Doing so facilitates a visualisation of the attainable performance of the process, subject to the rules imposed in generating the set of profiles. Dynamic optimisation can be implemented for temperature profile manipulation in order to investigate the possibility of the process benefitting from non-isothermal operation for specific production objectives.

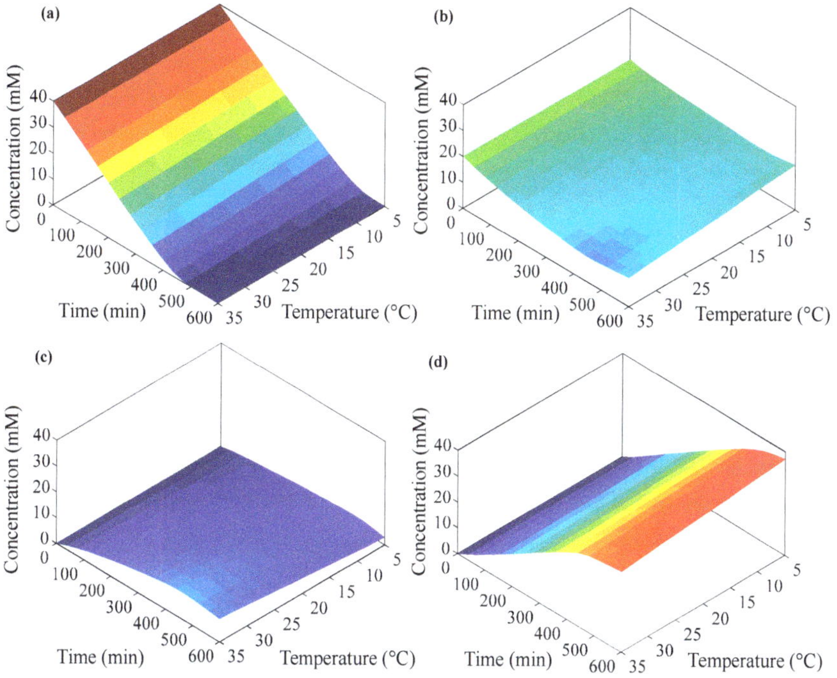

Figure 5. Isothermal concentration response surfaces as a function of batch time and operating temperature for (**a**) PHPGME, (**b**) 6-APA, (**c**) AMOX and (**d**) PHPG.

3.3. Non-Isothermal Dynamic Optimisation

The resulting temperature (control) and species concentration (state) trajectories for different product amoxicillin constraints imposed on the dynamic optimisation problem are shown in Figures 8 and 9, respectively. The temperature profiles varied significantly as the constraint concentration of amoxicillin in the reactor product mixture increased.

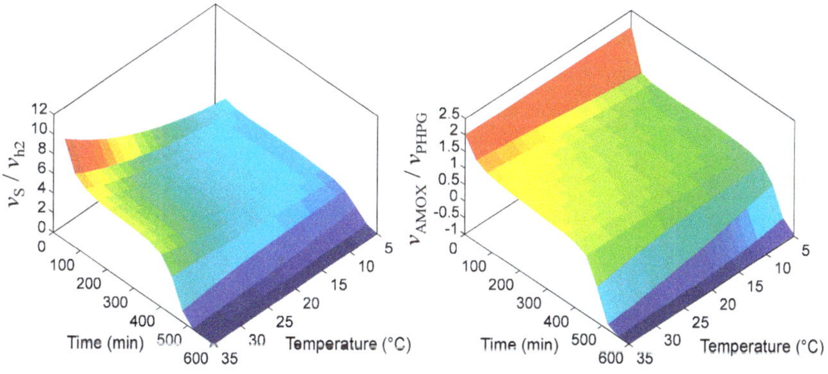

Figure 6. Isothermal response surface of reaction rate ratios as a function of batch time and temperature.

When the optimisation problem was unconstrained, the temperature gradually decreased from 296 K at t = 0 min to 278.15 K at t = 325 min (lower temperature bound), after which isothermal operation at this temperature continued for the remainder of the batch time. For C_{AMOX} (t_f) = 5 and

6 mM, the temperature profile decreased until t = 375 min and then increased again, with a higher final operating temperature required for a higher concentration constraint. When the constraint was set to C_{AMOX} (t_f) = 7 mM, a gradual decrease from 296 to 290 K over the batch duration was implemented. For C_{AMOX} (t_f) ≥ 8 mM, temperature profiles increased over the batch duration. At C_{AMOX} (t_f) = 9 mM, a steady increase from 295 to 305 K was observed. For C_{AMOX} (t_f) = 10 mM, the temperature increased sharply from 292 to 308 K (upper temperature bound) at t = 100 min. For C_{AMOX} (t_f) = 11 mM, the temperature increase rate was more drastic, increasing to the upper bound at t = 25 min, with subsequent isothermal operation for the remainder of the batch duration.

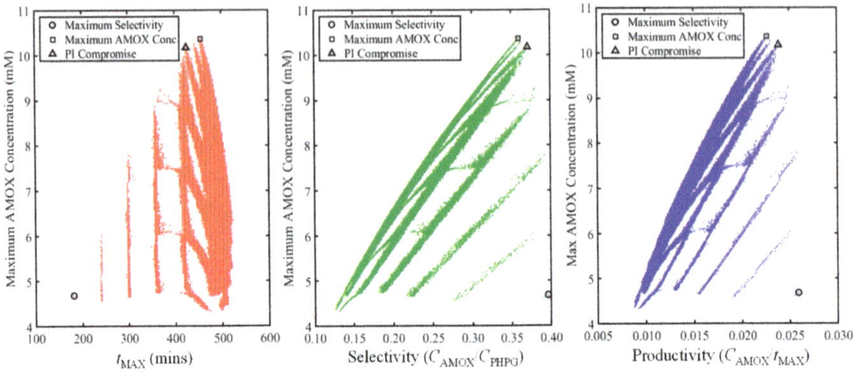

Figure 7. Non-isothermal simulation performance indices.

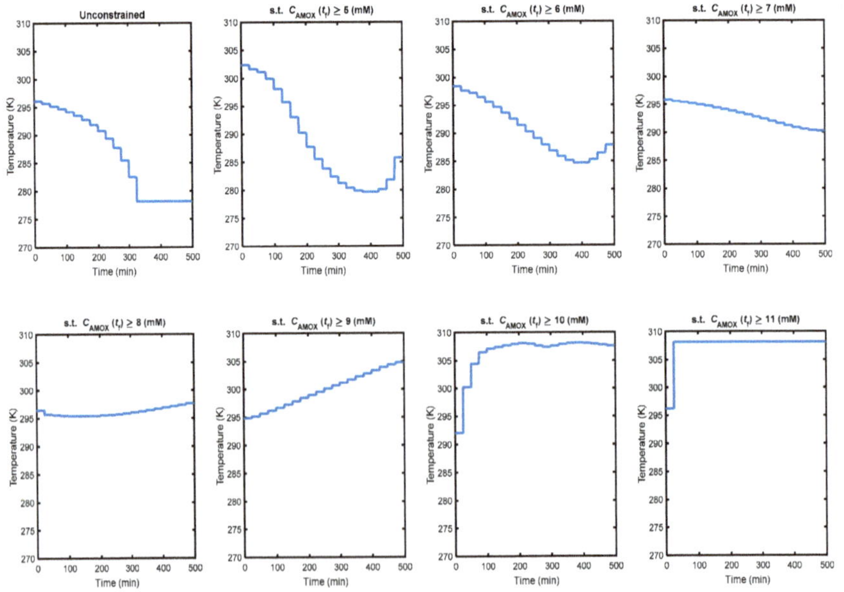

Figure 8. Optimal temperature trajectories for different production constraints.

These trends confirmed that lower amoxicillin target constraints required lower operating temperatures, with higher temperatures required for higher target concentrations. This behaviour was due to the effect of increasing temperatures favouring amoxicillin synthesis over undesired side hydrolysis, consistent with the system behaviour shown earlier in Figure 5. For the unconstrained case, the temperature was eventually pushed to the lower bound, and as the target amoxicillin concentration increased, the temperature was pushed to the upper bound. The temperature bounds imposed on the dynamic optimisation problem corresponding to the lower and upper temperature values at which experimental concentration data for kinetic and Arrhenius parameter regression were implemented [22]. Relaxing these bounds may have resulted in different optimal control trajectories. Industrial temperature controllers typically operate similarly to piecewise constant or linear profile behaviours, but there is inevitably a temperature lag between the constant temperature discretisations. The incorporation of time lag into temperature control will certainly affect the presented results: Such an analysis was outside the scope of this work, but may enhance our understanding of attainable optima for this work and other industrially relevant design cases. The optimal temperature profiles for each case presented are those attained from our optimisation, having initialised the solver with the same starting solution (isothermal at T = 295 K). There were different profiles that could still meet the same amoxicillin product concentration constraints (i.e., different routes to the same product specification). The isothermal concentration response surfaces in Figure 5 were unable to depict state evolution under non-isothermal reactor operation, and thus dynamic optimisation was implemented.

The resulting species concentration (state) trajectories are shown in Figure 9. In all cases, as substrates PHPGME and 6-APA were consumed, product amoxicillin and by-product PHPG were formed. For the unconstrained and lower amoxicillin concentration constraints (e.g., C_{AMOX} (t_f) = 5–7 mM), the target amoxicillin concentration was met before the concentration of 6-APA (i.e., the objective function) was maximised. This indicated that there was scope to also optimise the batch runtime in the dynamic optimisation problem as well, which can be considered in future work. The final species concentrations at the end of the batch runtime (t_f = 500 min) are shown in Figure 10. As the product amoxicillin constraint was increased, the maximum objective function value decreased.

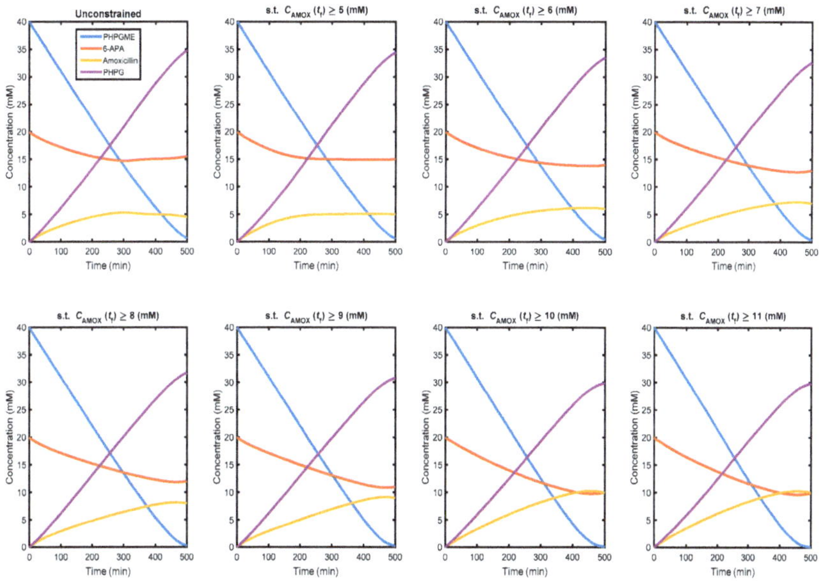

Figure 9. Optimal concentration trajectories for different production constraints.

The objective of the dynamic optimisation was to minimise 6-APA consumption (posed by maximising the final 6-APA concentration at the end of the batch duration). Figure 10 shows the fraction of 6-APA from the process feed that was present in the batch product versus that consumed in the reaction network. In all design cases (varying the amoxicillin product concentration constraint), a significant portion of the fed 6-APA was present in the product mixture.

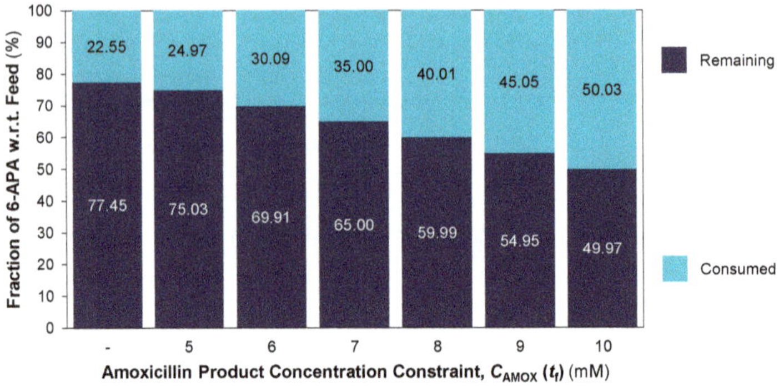

Figure 10. Fraction of 6-APA remaining in the batch product versus that consumed via reaction.

It can be seen from Figure 9 that when C_{AMOX} (t_f) = 11 mM was specified, the constraint was not met. Consideration of the maximum attainable objective function values for a broad range of C_{AMOX} (t_f) (Figure 11) explains this result. This was a result of there being a maximum product concentration of amoxicillin attainable from the given feed substrate concentrations. Figure 12 plots the attained maximised objective function for different imposed amoxicillin product concentration constraints. Until a certain amoxicillin product concentration (Point A), the maximum attainable objective function value was the same: This was the maximum attainable value from the given initial conditions (feed concentrations). As the product amoxicillin concentration increased, the maximum attainable objective function decreased until the maximum constraint, C_{AMOX} (t_f) = 9.927 mM, allowing for an objective function, $C_{6\text{-}APA}$ (t_f) = 10.067 mM (Point B). The limiting points (A and B) highlighted in Figure 12 resulted from the initial concentration conditions of the system.

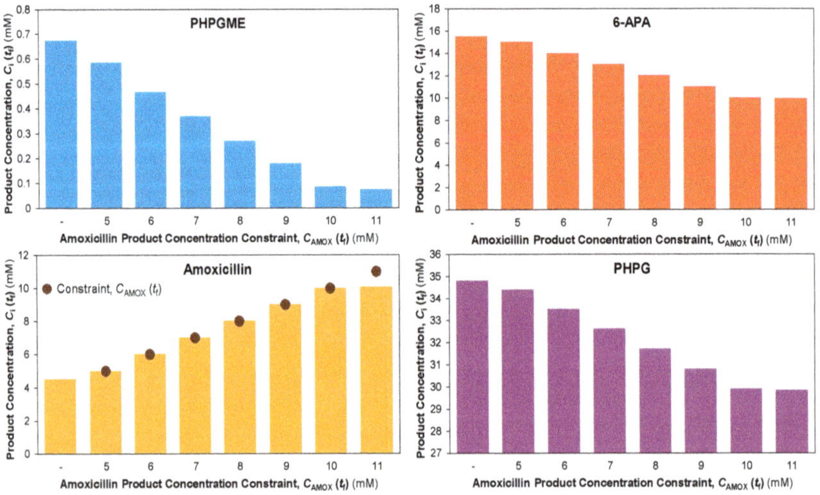

Figure 11. Final concentrations (colour scheme congruent with Figure 9 trajectories).

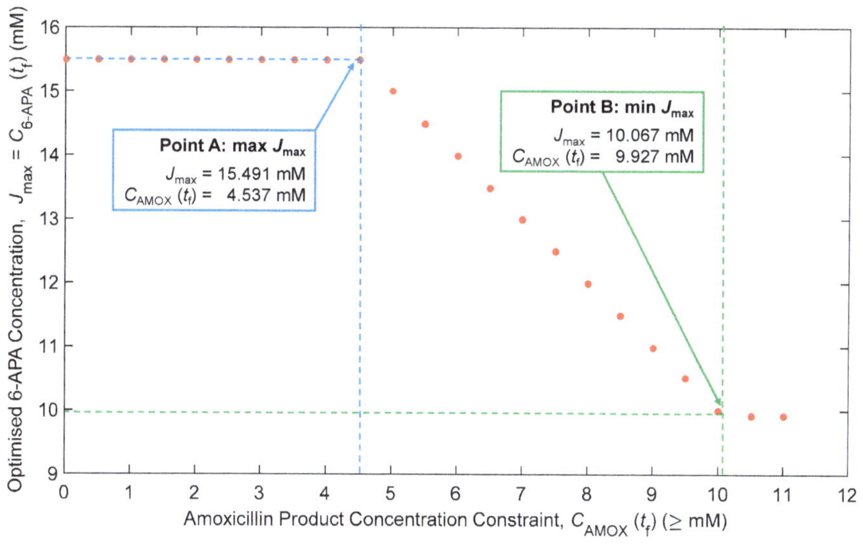

Figure 12. Maximum 6-APA concentration for different constrained amoxicillin concentrations.

The initial concentration values of PHPGME and 6-APA used in this work were the same as those implemented in the literature, from which experimental data were used for parameter regression. These values corresponded to typical industrial application ranges [20]. In the literature, it was observed that concentrations of 6-APA > 20 mM led to a slower formation of the acyl-enzyme complex in the amoxicillin synthesis reaction, which was detrimental to process performance [20,22]. For this reason, the initial conditions (i.e., reagent concentrations) considered in this work were kept as demonstrated in the literature. Additionally, an investigation of the applicability of the model at higher reactor volumes for increased production scales would be of value, requiring further experimental validation and corroboration. The results presented in this work assumed perfect mixing and heat transfer in the batch reactor. As the process is scaled up, deviations from this ideal behaviour assumption must be considered when interpreting the results presented in this work.

Measuring the material efficiency of different design cases is also an important consideration in pharmaceutical production [31]. A variety of green chemistry metrics exist for quantification of the effectiveness of processes, the applicability of which depends on the process [32]. Here, we compare the efficiency of different optimisation scenarios (varying $C_{AMOX}(t_f)$ considerations) via the reaction mass efficiency (RME), calculated by Equation (19).

$$\text{RME} = \frac{m_{\text{AMOX}}(t_f)}{m_{\text{PHPGME}}(t_0) + m_{6-\text{APA}}(t_0)} \tag{19}$$

The RME is calculated as the mass ratio of amoxicillin (m_{AMOX}) at the end of a batch run to the masses of starting materials (PHPGME (m_{PHPGME}) and 6-APA ($m_{6\text{-APA}}$)) at the start ($t_0 = 0$). Values of RME for different product amoxicillin concentration constraints are shown in Figure 13. As the specified amoxicillin concentration constraint increases, the RME increases, which is expected. All values of RME were relatively midrange with respect to typical pharmaceutical manufacturing processes [33]. The effect of scale-up on material efficiencies, as well as plant-wide costs, is an important consideration during process design and optimisation studies such as this one [34].

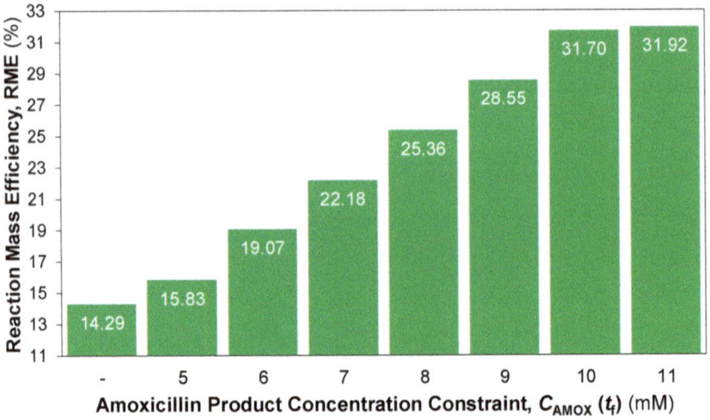

Figure 13. Reaction mass efficiencies (RMEs) for different constrained amoxicillin concentrations.

Figure 14 shows purity, with different species as the reference point corresponding to optima (maximum product 6-APA concentrations) for different product amoxicillin concentration constraints, calculated from the final concentration values shown in Figure 10. In all cases, PHPGME and 6-APA (reagent) concentrations decreased as amoxicillin purities increased. While PHPG contents decreased, they remained high compared to other species in the mixture. A consideration of mixture compositions following synthesis is important for purification and separation process design, particularly for crystallisation design [35]. Understanding the partitioning of different species between liquid phase mixtures and solid phase product crystals is essential for understanding the effects of different operating and design parameters on crystal product attributes (purity, size distribution, etc.). Another method of circumventing the accumulation of impurities in crystals is reactive crystallisation, which aims to preferentially crystallise the desired compound from the reaction mixture solution: This has been implemented for various β-lactam antibiotics [16,17]. While such methods may allow for improved purities over traditional crystallisation, ensuring rates of reaction compared to mass transfer rates to avoid high supersaturation is paramount, as this can lead to undesirably wide crystal size distributions and low mean sizes.

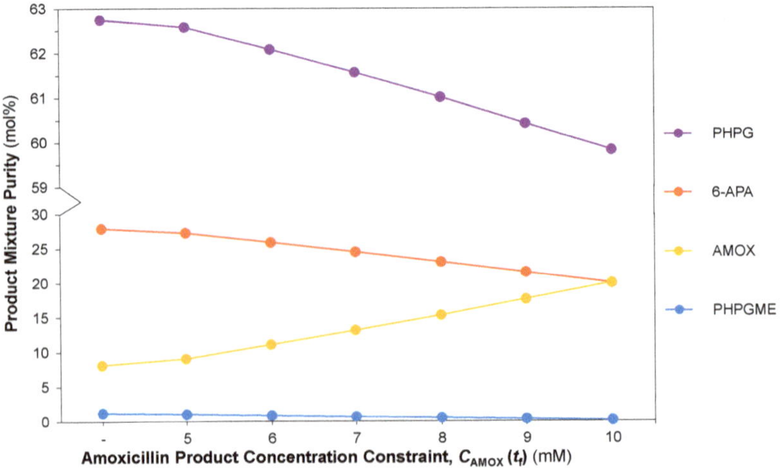

Figure 14. Product mixture content for different amoxicillin product concentration constraints (colour scheme congruent with Figures 9 and 11).

The basis of the dynamic optimisation cases considered in this work was to establish optimal temperature profiles to meet specific amoxicillin product concentrations while minimising the consumption of key feedstock (6-APA). Varying constraints on the amoxicillin product concentrations were considered to investigate the effect of different desired mixture qualities on optimal temperature control trajectories. The objective of the dynamic optimisation problem was not to maximise amoxicillin production, but rather to produce at least a certain desired quantity of amoxicillin while minimising consumption of the key reagent 6-APA. In the presented results, dynamic temperature profiles were confirmed to be optimal for attaining a range of target amoxicillin concentrations while minimising 6-APA consumption. The highest target amoxicillin concentration considered had a nearly isothermal temperature profile over the batch duration: If the objective of the optimisation had been simply the maximisation of amoxicillin product concentration, isothermal profiles would likely have been preferable. However, for certain desired amoxicillin product concentrations, dynamic temperature profiles were confirmed to be optimal.

4. Conclusions

A kinetic model for the batch enzymatic synthesis of amoxicillin was considered in this work for dynamic modelling and optimisation. The novelty of this work is in the following:

1. The introduction of temperature dependency to the implemented existing kinetic model [20] using experimental concentration data [22] for the batch enzymatic synthesis of amoxicillin; and
2. The first implementation of dynamic temperature profile optimisation in order to meet specific product quality constraints and minimise feedstock consumption in β-lactam antibiotic production.

Temperature dependency in the model was introduced through the regression of Arrhenius parameters to allow for non-isothermal modelling and optimisation of the temperature trajectories to minimise key substrate (6-APA) consumption subject to varying constraints on the product amoxicillin concentration. The attainable minimal substrate consumption was strongly dependent on the amoxicillin product concentration. This work presents the first batch reactor temperature trajectory optimisation for amoxicillin synthesis and can be extended to other antibiotics given similar reaction networks and the availability of model kinetic parameters.

Author Contributions: Conceptualisation, A.B.C., A.D.R., S.D. and D.I.G.; methodology, A.D.R., A.B.C., S.D. and D.I.G.; software, A.B.C. and A.D.R.; validation, A.B.C. and A.D.R.; formal analysis, A.B.C., A.D.R. and S.D.; writing, A.B.C., A.D.R., S.D. and D.I.G.; supervision, D.I.G.

Funding: A.B.C. acknowledges the financial support of an Engineering and Physical Sciences Research Council (EPSRC) Vacation Bursary. A.D.R. acknowledges the Eric Birse Charitable Trust for a Doctoral Fellowship. S.D. acknowledges the financial support of the Engineering and Physical Sciences Research Council (EPSRC) via a Doctoral Training Partnership (DTP) PhD Fellowship. D.I.G. acknowledges a Royal Academy of Engineering (RAEng) Industrial Fellowship.

Conflicts of Interest: The authors declare no conflicts of interest. The funders had no role in the study. Tabulated and cited literature data suffice for the reproduction of all original simulation and optimisation results, and no other supporting data are required to ensure reproducibility.

References

1. Zaffiri, L.; Gardner, J.; Toledo-Pereyra, L.H. History of antibiotics. from salvarsan to cephalosporins. *J. Investig. Surg.* **2012**, *25*, 67–77. [CrossRef] [PubMed]
2. Balkhi, B.; Araujo-Lama, L.; Seoane-Vazquez, E.; Rodriguez-Monguio, R.; Szeinbach, S.L.; Fox, E.R. Shortages of systemic antibiotics in the USA: how long can we wait? *J. Pharm. Heal. Serv. Res.* **2013**, *4*, 13–17. [CrossRef]
3. Pulcini, C.; Beovic, B.; Béraud, G.; Carlet, J.; Cars, O.; Howard, P.; Levy-Hara, G.; Li, G.; Nathwani, D.; Roblot, F.; et al. Ensuring universal access to old antibiotics: a critical but neglected priority. *Clin. Microbiol. Infect.* **2017**, *23*, 590–592. [CrossRef] [PubMed]
4. Russell, M.G.; Jamison, T.F. Seven-step continuous flow synthesis of linezolid without intermediate purification. *Angew. Chemie Int. Ed.* **2019**. [CrossRef]

5. Lin, H.; Dai, C.; Jamison, T.F.; Jensen, K.F. A rapid total synthesis of ciprofloxacin hydrochloride in continuous flow. *Angew. Chemie Int. Ed.* **2017**, *56*, 8870–8873. [CrossRef] [PubMed]
6. Giordano, R.C.; Ribeiro, M.P.A.; Giordano, R.L.C. Kinetics of β-lactam antibiotics synthesis by penicillin G acylase (PGA) from the viewpoint of the industrial enzymatic reactor optimization. *Biotechnol. Adv.* **2006**, *24*, 27–41. [CrossRef] [PubMed]
7. Food and Drug Administration (FDA). Sales of antibacterial drugs in kilograms. 2010; 4–6.
8. Food and Drug Administration (FDA). Sales of antibacterial drugs in kilograms. 2012; 5–8.
9. Hamed, R.B.; Gomez-Castellanos, J.R.; Henry, L.; Ducho, C.; McDonough, M.A.; Schofield, C.J. The enzymes of β-lactam biosynthesis. *Nat. Prod. Rep.* **2013**, *30*, 21–107. [CrossRef]
10. Elander, R.P. Industrial production of β-lactam antibiotics. *Appl. Microbiol. Biotechnol.* **2003**, *61*, 385–392. [CrossRef]
11. Laxminarayan, R. The state of the world's antibiotics in 2018. 3 July 2018.
12. Gerogiorgis, D.I.; Jolliffe, H.G. Continuous pharmaceutical process engineering and economics investigating technical efficiency, environmental impact and economic viability. *Chem. Today* **2015**, *33*, 29–32.
13. Diab, S.; Gerogiorgis, D.I. Process modelling, simulation and technoeconomic evaluation of crystallisation antisolvents for the continuous pharmaceutical manufacturing of rufinamide. *Comput. Chem. Eng.* **2018**, *111*, 102–114. [CrossRef]
14. Rodman, A.D.; Gerogiorgis, D.I. An investigation of initialisation strategies for dynamic temperature optimisation in beer fermentation. *Comput. Chem. Eng.* **2019**, *124*, 43–61. [CrossRef]
15. Rodman, A.D.; Gerogiorgis, D.I. Dynamic optimization of beer fermentation: sensitivity analysis of attainable performance vs. product flavour constraints. *Comput. Chem. Eng.* **2017**, *106*, 582–595. [CrossRef]
16. McDonald, M.A.; Bommarius, A.S.; Rousseau, R.W.; Grover, M.A. Continuous reactive crystallization of β-lactam antibiotics catalyzed by penicillin G acylase. part I: model development. *Comput. Chem. Eng.* **2019**, *123*, 331–343. [CrossRef]
17. McDonald, M.A.; Bommarius, A.S.; Grover, M.A.; Rousseau, R.W. Continuous reactive crystallization of β-lactam antibiotics catalyzed by penicillin G acylase. part II: case study on ampicillin and product purity. *Comput. Chem. Eng.* **2019**, *126*, 332–341. [CrossRef]
18. Encarnación-Gómez, L.G.; Bommarius, A.S.; Rousseau, R.W. Crystallization kinetics of ampicillin using online monitoring tools and robust parameter estimation. *Ind. Eng. Chem. Res.* **2016**, *55*, 2153–2162. [CrossRef]
19. McDonald, M.A.; Bommarius, A.S.; Rousseau, R.W. Enzymatic reactive crystallization for improving ampicillin synthesis. *Chem. Eng. Sci.* **2017**, *165*, 81–88. [CrossRef]
20. Gonçalves, L.R.B.; Fernández-Lafuente, R.; Guisán, J.M.; Giordano, R.L.C. The role of 6-aminopenicillanic acid on the kinetics of amoxicillin enzymatic synthesis catalyzed by penicillin G acylase immobilized onto glyoxyl-agarose. *Enzyme Microb. Technol.* **2002**, *31*, 464–471. [CrossRef]
21. UK Department of Health. Antimicrobial resistance empirical and statistical evidence-base. 2016.
22. Alemzadeh, I.; Borghei, G.; Va, L.; Roostaazad, R. Enzymatic synthesis of amoxicillin with immobilized penicillin G acylase. *Trans. C Chem. Chem. Eng.* **2010**, *17*, 106–113.
23. Schroën, C.G.P.H.; Van Roon, J.L.; Beefink, H.H.; Tramper, J.; Boom, R.M. Membrane applications for antibiotics production. *Desalination* **2009**, *236*, 78–84. [CrossRef]
24. Fodi, T.; Didaskalou, C.; Kupai, J.; Balogh, G.T.; Huszthy, P.; Szekely, G. Nanofiltration-enabled in situ solvent and reagent recycle for sustainable continuous-flow synthesis. *ChemSusChem* **2017**, *10*, 3435–3444. [CrossRef] [PubMed]
25. Pereira, S.C.; Castral, T.C.; Ribeiro, M.P.A.; Giordano, R.L.C.; Giordano, R.C. Green route for amoxicillin production through the integration with the recycle of the by-product (p-hydroxyphenylglycine). In Proceedings of the Brazilian Congress of Chemical Engineering, São Paulo, Spain, 7–12 June 2015; Blücher, E., Ed.; pp. 1823–1830.
26. Marler, R.T.; Arora, J.S. Survey of multi-objective optimization methods for engineering. *Struct. Multidiscip. Optim.* **2004**, *26*, 369–395. [CrossRef]
27. Cuthrell, J.E.; Biegler, L.T. On the optimization of differential-algebraic process systems. *AIChE J.* **1987**, *33*, 1257–1270. [CrossRef]
28. Logsdon, J.S.; Biegler, L.T. Accurate solution of differential-algebraic optimization problems. *Ind. Eng. Chem. Res.* **1989**, *28*, 1628–1639. [CrossRef]

29. Wächter, A.; Biegler, L.T. On the implementation of an interior-point filter line-search algorithm for large-scale nonlinear programming. *Math. Program.* **2006**, *106*, 25–57. [CrossRef]
30. Čižniar, M.; Fikar, M.; Latifi, M.A. MATLAB dynamic optimisation code DynOpt. User's guide, technical report. *KIRP FCHPT STU Bratislava* **2005**.
31. Hicks, M.B.; Farrell, W.; Aurigemma, C.; Lehmann, L.; Weisel, L.; Nadeau, K.; Lee, H.; Moraff, C.; Wong, M.; Huang, Y.; et al. Making the move towards modernized greener separations: introduction of the analytical method greenness score (AMGS) calculator. *Green Chem.* **2019**, *21*, 1816–1826. [CrossRef]
32. Sheldon, R.A. Fundamentals of green chemistry: Efficiency in reaction design. *Chem. Soc. Rev.* **2012**, *41*, 1437–1451. [CrossRef] [PubMed]
33. Ribeiro, M.G.T.C.; Machado, A.A.S.C. Greenness of chemical reactions—limitations of mass metrics. *Green Chem. Lett. Rev.* **2013**, *6*, 1–18. [CrossRef]
34. Jolliffe, H.G.; Gerogiorgis, D.I. Plantwide Design and Economic Evaluation of Two Continuous Pharmaceutical Manufacturing (CPM) Cases: Ibuprofen and Artemisinin. *Comput. Aided Chem. Eng.* **2015**, *37*, 2213–2218.
35. Li, J.; Lai, T.C.; Trout, B.L.; Myerson, A.S. Continuous crystallization of cyclosporine: the effect of operating conditions on yield and purity. *Cryst. Growth Des.* **2017**, *17*, 1000–1007. [CrossRef]

processes

MDPI

Article

Global Sensitivity Analysis of a Spray Drying Process

Satyajeet Bhonsale †, Carlos André Muñoz López † and Jan Van Impe *

Chemical and BioProcess Technology and Control, Department of Chemical Engineering, KU Leuven, Gebroeders de Smetstraat 1, 9000 Ghent, Belgium

* Correspondence: jan.vanimpe@kuleuven.be; Tel.: +32-477-256-172

† These authors contributed equally to this work.

Received: 23 July 2019; Accepted: 16 August 2019; Published: 23 August 2019

Abstract: Spray drying is a key unit operation used to achieve particulate products of required properties. Despite its widespread use, the product and process design, as well as the process control remain highly empirical and depend on trial and error experiments. Studying the effect of operational parameters experimentally is tedious, time consuming, and expensive. In this paper, we carry out a model-based global sensitivity analysis (GSA) of the process. Such an exercise allows us to quantify the impact of different process parameters, many of which interact with each other, on the product properties and conditions that have an impact on the functionality of the final drug product. Moreover, classical sensitivity analysis using the Sobol-based sensitivity indices was supplemented by a polynomial chaos-based sensitivity analysis, which proved to be an efficient method to reduce the computational cost of the GSA. The results obtained demonstrate the different response dependencies of the studied variables, which helps to identify possible control strategies that can result in major robustness for the spray drying process.

Keywords: quality by design; pharmaceutical manufacturing; polynomial chaos; global sensitivity analysis; spray drying

1. Introduction

Spray drying is a widely-used unit operation in the production of high-value-added products in the food, fertilizers, chemical, and pharmaceutical industries [1–4]. Its application ranges from the production of milk and other diary products, to the very complex formulations of composite materials used in medicines, to biological products for which few other drying technologies are feasible. Spray drying is a cost-efficient, continuous, and relatively easy to operate process [5]. A typical spray dryer, as shown in Figure 1, consists of two sections, an atomization section and a drying section. The atomizer disperses the liquid mixture into small droplets, which then fall into the drying chamber. Small droplets dispersed in a hot gas medium offer large surface areas for mass and energy transfer, while guaranteeing short exposure of the product to high temperatures [6,7]. The dried product is then collected using an array of cyclones and filters. Spray drying allows a more specific functional design of products. It provides high flexibility on the input mixture because the liquid feed can be provided in various forms, e.g., solutions, emulsions, or suspensions. The nature of the particulate product ranges from pure substances, solid blends, to composite materials, and viable biological components. With proper tuning, spray drying offers relatively narrow particle size distributions in the range between nano- to micro-sized particles [8–10], and good control over other particle properties like residual solvent content, morphology, and density. The major tuning parameters include the flow rate and temperature of the feed and drying gas;

concentration, surface tension, viscosity, and density of the liquid feed; flow configuration; and atomizer geometry. Any variations in these process conditions or the input feed properties will thus affect the downstream properties.

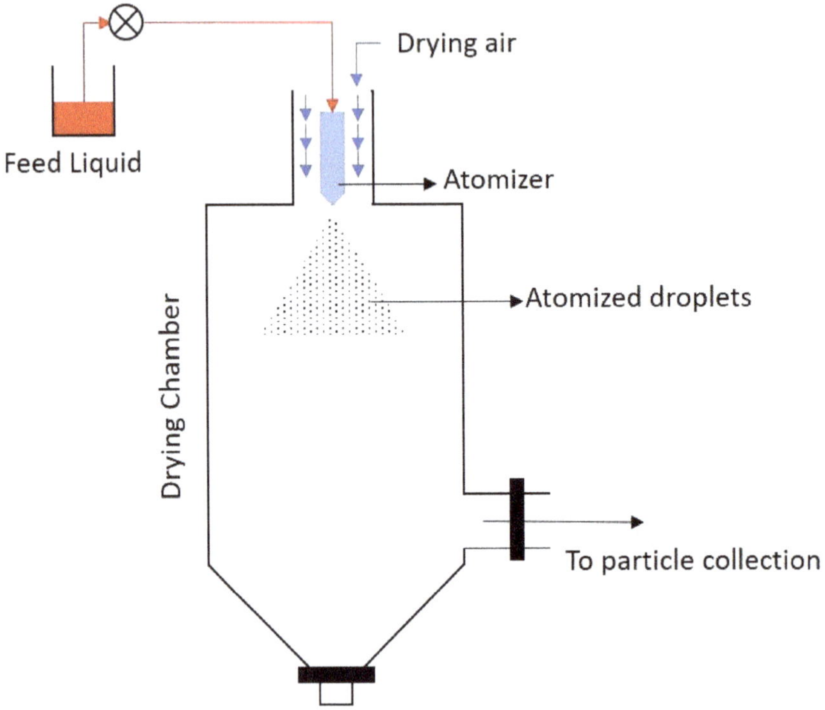

Figure 1. Schematic of a spray dryer.

The continuous push from regulatory agencies, the tightening quality requirements, and the increase in the number of specialized drug products drive the research on spray drying towards the development of model-based approaches that replace the traditional empirical practices. Similar to many other unit operations that handle particulate systems, the setup of large-scale spray drying operations remains highly empirical [1]. Due to the complexity of the multiphase system involved, the process conditions for spray drying are traditionally determined by trial and error experiments. This approach normally involves large investments in time and money to determine those process conditions that result in a dried product with the desired properties. In the pharmaceutical industry, the quality by design paradigm (QbD) has already pushed for a more systematic view of the exploitation of the experimental knowledge [11]. Applying QbD to the spray drying process entails: (i) determination of powder properties, which are critical quality attributes (CQAs) based on the impact they have on the performance of the drug product, (ii) identification of the critical process parameters (CPPs) and input material attributes (CMAs) that affect the CQAs, and (iii) establishment of a design space and control strategy in terms of CPPs and CMAs that can guarantee the CQAs' requirements.

Within this context, this work evaluates the use of global sensitivity analysis (GSA) as a model-based approach to apply the principles of QbD to the spray drying process. This approach bridges the research drive towards model-based methods for spray drying and the goals of QbD in the pharmaceutical industry. Previous research in these lines has focused on developing models for spray drying [1,12–16] or on QbD for spray drying using the design of experiments and defining the design space based on statistical analysis of the experimental data [17–20]. In this work, the focus is put first on the evaluation of the methods to apply GSA to the spray drying process, followed by the critical analysis of the results to demonstrate its potential to exploit the knowledge of the system contained in the model. GSA offers a structured approach to identify, quantify, and rank the influences of the CPPs and CMAs on the CQAs.

Several models have been developed in the past for the spray drying process. The level of detail and associated complexity of these varies from considering one dimensional differential equations [12,13] to complex partial differential equations, which aim to capture gradients in more multiple dimensions [15]. CFD simulations show the complex flow patterns that are developed inside the drying chamber due to the geometry of the chamber, velocity and pressure gradients in the gas, as well as the size distribution of the particles [16]. Studies on atomization generally consider empirical equations to model specific types of nozzles. Since the aim of this work is to demonstrate the power of GSA to capture the global relation between variables and the mean responses of the CQAs, a validated one-dimensional system of differential equations is chosen to describe the spray drying process [1]. This model is described in Section 2.1. Additionally, two different atomization nozzle configurations are considered to evaluate the differences induced in the process and the product. The pneumatic and the pressure nozzle considered are discussed in Section 2.3.

A traditional approach to GSA is the use of Sobol sensitivity indices, which result from an analysis of variance on the system [21]. This method is discussed in Section 2.2.1. An analytical solution to partial variances is almost impossible for complex mathematical models, and typically, approximations based on (quasi) Monte Carlo sampling are used. The current best practices to calculate the sensitivity indices make use of Saltelli's approach with a sufficient number of quasi-random samples [22,23]. However, a large number of model evaluations for complex models lead to exorbitant computational times. More computationally-efficient methods for the approximation of sensitivity indices are based on surrogate meta-models. These approaches have already been investigated for the GSA of (bio)chemical processes [24]. In this work, polynomial chaos expansion (PCE) is used as a way to construct the meta-model for spray drying and reduce the computational burden for the GSA [25–27]. PCE is based on the probabilistic projection of the model output on the basis of orthogonal stochastic polynomials. In this work, arbitrary or data-driven is investigated, as it does not depend on an assumption of any specific probability distribution functions (PDF) for the inputs. To assess the PCE-based sensitivity indices, we assume that the indices based on Saltelli's approach are true iff the number of samples is large enough. The question of the size of the sample set is discussed in Section 3.1. In the same section, the sensitivity indices calculated using the PCE are then compared to the Saltelli approximations.

The results obtained are discussed in Sections 3.2 and 3.3. Two main aspects are reviewed based on these results. (i) The validity of GSA using arbitrary or data-driven polynomial chaos expansions (aPCE) is evaluated with respect to the indices computed from the comprehensive variance analysis based on Saltelli's method. (ii) A critical discussion is provided on the results of the GSA for the spray drying process.

2. Materials and Methods

2.1. Spray Drying Model

Considering the co-current gas flow spray dryer, the process proceeds in the following stages. First, the liquid feed is atomized using special nozzles, and the droplets enter the drying chamber. The solvent is then evaporated from the droplets by the drying gas fed to the chamber. At one point along the drying chamber, the droplets transition to solid particles, which then need to be separated from the drying gas. In this paper, the atomization section is modeled independently of the drying section. The spray drying model developed in [1] was used for the study. The spray dryer is modelled axially, ignoring any radial effects under a number of other simplifying assumptions.

For the solids, the humidity (W_p) in the droplet/particle follows Equation (1)

$$\frac{dW_p}{dz} = -\frac{\pi d_p \dot{m}_v}{v_p m_s}, \quad W_p(z=0) = W_{p0} \tag{1}$$

where z is the axial distance, v_p and d_p are the particle velocity and diameter, m_s is the mass of solids in the droplet (estimated from the initial liquid concentration), and $\dot{m}_v$ is the rate of evaporation. The particle humidity evolves until an equilibrium moisture content is reached. After this point, it stabilizes, and the evaporation rate drops to zero.

A particle/droplet in the spray dryer goes through various stages of drying. Initially, as the rate of evaporation is balanced by the rate of moisture transfer from the center of the particle to the surface, the droplet shrinks in size, while its density increases. Further down the chamber, a critical moisture content is reached where the droplet transitions to being a "wet particle", and the size of the particle remains constant. The transition of the wet particle to a dry particle is limited by the rate of moisture being wicked to the surface of the particle. Once the equilibrium moisture content is reached, the particle stops drying, and its density (ρ_p) and size (d_p) remain constant. Cotabarren et al. [1] modeled the process as follows.

When $W_p \geq W_{pc}$,

$$\frac{dd_p}{dz} = \frac{d_{p0}\dot{m}_v \pi d_p^2}{3 m_s v_p}\left(\frac{\rho_{p0}-\rho_w}{\rho_p-\rho_w}\right)^{-\frac{2}{3}}\left(\frac{\rho_{p0}-\rho_w}{(\rho_p-\rho_w)^2}\right)\left(\frac{1-\frac{\rho_s}{\rho_w}}{(1+\frac{\rho_s}{\rho_w}W_p)^2}\right), \quad d_p(z=0) = d_{p0} \tag{2}$$

$$\frac{d\rho_p}{dz} = -\frac{\dot{m}_v \pi d_p^2 \rho_s}{m_s v_p}\frac{1-\frac{\rho_s}{\rho_w}}{(1+\frac{rho_s}{rho_w}W_p)^2}, \quad \rho_p(z=0) = \rho_{p0} \tag{3}$$

when $W_{peq} \leq W_p < W_{pc}$:

$$\frac{dd_p}{dz} = 0 \tag{4}$$

$$\frac{d\rho_p}{dz} = -\frac{6\dot{m}_v}{d_p v_p} \tag{5}$$

and when $W_p < W_{peq}$,

$$\frac{d\rho_p}{dz} = 0 \tag{6}$$

The particle velocity is modeled by balancing the gravity with the drag forces acting on the particles.

$$\frac{dv_p}{dz} = g\left(\frac{\rho_p - \rho_a}{\rho_p v_p}\right)^{-\frac{3}{4}} - \frac{3C_D\rho_a}{4d_p\rho_p v_p}(v_a - v_p)|v_a - v_p| \quad (7)$$

The rate of evaporation is described as a function of relative humidity of the gas and its saturation moisture content at the particle surface temperature [28].

$$\dot{m}_v = \beta(Y_{sat} - Y_b) \quad (8)$$

$$Y_{sat} = \frac{P_v\tilde{M}_w}{(P - P_v)\tilde{M}_a} \quad (9)$$

The vapor pressure P_v was calculated using Antoine's equations. Pis the atmospheric pressure, and $\tilde{M}_{\{a,w\}}$ is the molecular weight of the gas (air) and solvent (water). The mass transfer coefficient is estimated as:

$$\beta = \frac{\alpha\rho_a\mathcal{D}_{eff}}{k_a} \quad (10)$$

The heat transfer coefficient α is estimated using empirical formulation [28,29]

$$Nu = 2 + 0.6Re^{0.5}Pr^{0.33} \quad (11)$$

where Nu is the Nusselt number, Re the particle Reynolds number, and Pr the Prandtl number. The energy balance on a single particle/droplet is given by:

$$\frac{dT_p}{dz} = \frac{\pi d_p^2[\alpha(T_a - T_p) - \dot{m}_v\Delta H_{ev}]}{v_p m_s(c_{p,s} + W_p c_{p,w})}, \quad T_p(z=0) = T_{p0} \quad (12)$$

where T_p is the particle temperature and ΔH_{ev} is the evaporation enthalpy of water.

The gas phase balances consider the total number of droplets, N_t, which is estimated as:

$$N_t = \frac{\dot{M}_l}{V_{p0}\rho_{p0}} \quad (13)$$

where $\dot{M}_l$ is the liquid mass flowing though the atomization nozzles and V_{p0} is the initial droplet volume.

The gas moisture (Y_b) balance is written as:

$$\frac{dY_b}{dz} = N_t\frac{\dot{m}_v\pi d_p^2}{\dot{M}_a v_p}, \quad Y_b(z=0) = Y_{b0} \quad (14)$$

The drying gas temperature (T_a) evolves as follows:

$$\frac{dT_a}{dz} = -\frac{N_t\pi d_p^2(\dot{m}_v c_{p,v} + \alpha)(T_a - T_p)}{v_p\dot{M}_a(c_{p,a} + X_b c_{p,v})} + \frac{U(T_a - T_{amb})\pi D_c}{\dot{M}_a(c_{p,a} + X_b c_{p,v})}, \quad T_a(z=0) = T_{a0} \quad (15)$$

Further details on the model can be found in [1]. The limits of the empirical relations used to calculate the dimensionless numbers (i.e., Re, Pr, Nu) and the mass and energy transfer coefficients are valid for the wide range of possible conditions observed in a spray drying chamber. These relations were tested in [29], in the range $0 \leq Re \leq 200$ and for gas temperatures up to 220 °C.

A typical spray drying profile obtained through the solution of the above model is depicted in Figure 2. The profiles observed for all variables illustrate that most of the changes occur in the first part of the drying chamber (i.e., $Z <\sim 0.14\,(\mathrm{m})$); this is before the equilibrium humidity is reached. After this point, most of the variables display a steady value. Only the temperatures continue to change due to the heat transfer through the exterior wall of the drying chamber. In this part, the temperature of the gas and of the particle are equal. In the section before the equilibrium point, the droplet/particle velocity determines two distinctive phases. At the very beginning, when the droplet velocity is significantly higher than that of the air (i.e., $Z <\sim 0.02\,(\mathrm{m})$), the conditions for mass transfer are favored by the turbulence around the droplets, and therefore, a high evaporation rate of the solvent can be observed. This causes the temperature of the droplet to drop because energy is consumed for evaporation faster than it is transferred from the hot air. The rate of change in droplet size and humidity is also distinctive in this section. Once the droplets have reached terminal velocity, the rate of evaporation is controlled by the energy transferred from the hot air to the droplet, and so the temperature stabilizes around a given value. In this section, the particle size and humidity continue to decrease. This phase ends when the particle has reached the equilibrium humidity. On the one hand, this stops the evaporation, reaching the final particle size, humidity, and density. The temperature, on the other hand, rises very quickly to reach equilibrium with the air temperature.

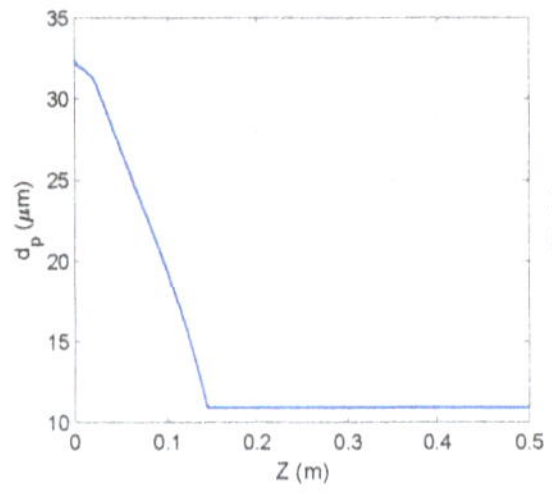

(a) Particle Diameter d_p

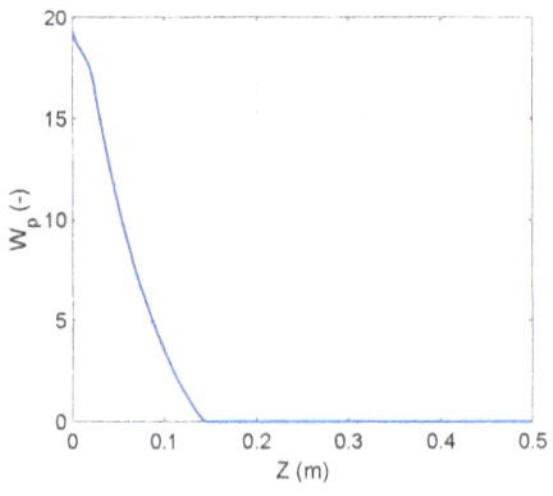

(b) Residual Solvent Content W_p

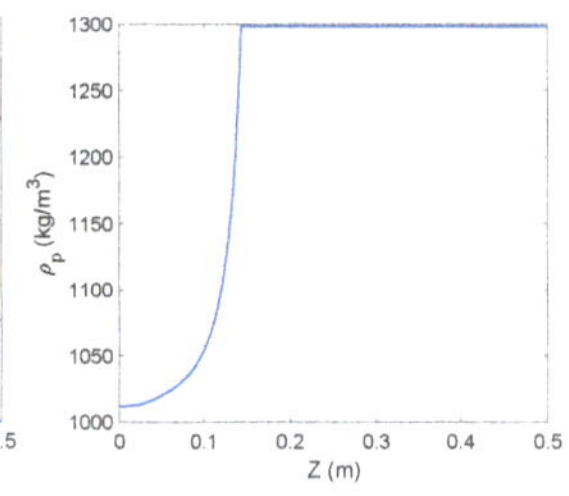

(c) Particle Density ρ_p

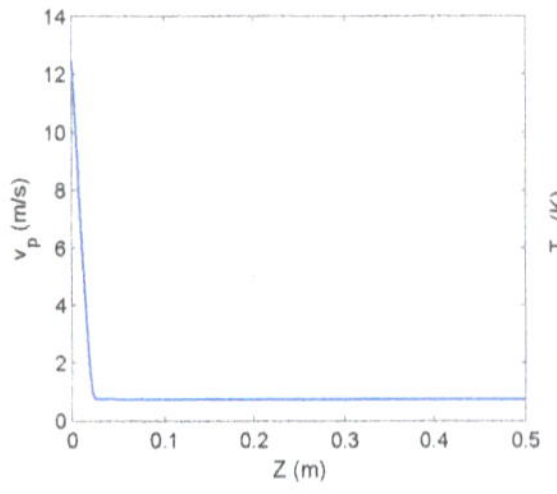

(d) Particle Velocity v_p

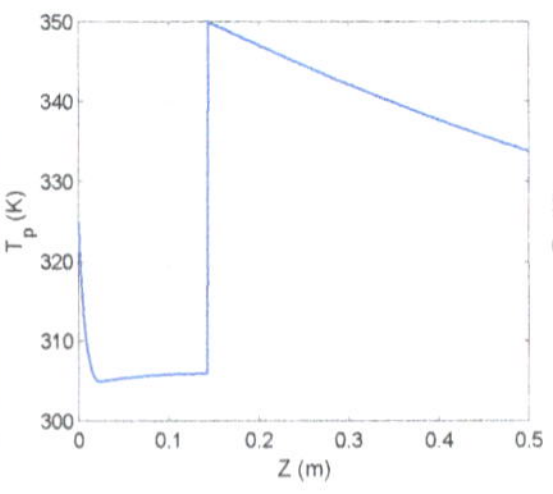

(e) Particle Temperature T_p

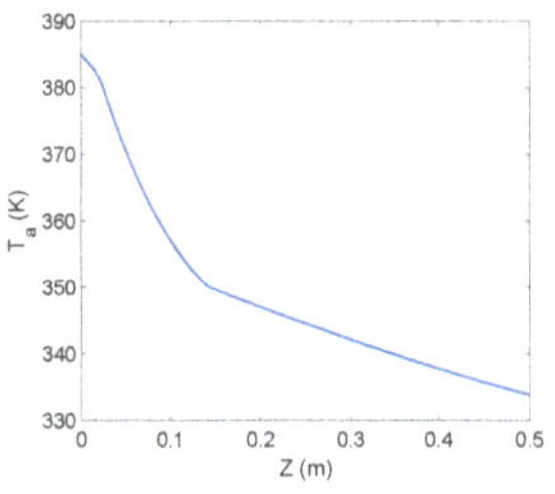

(f) Drying Air Temperature T_a

Figure 2. Example of the spray drying output evolution along the length of the drying chamber at nominal conditions.

2.2. Global Sensitivity Analysis

The aim of sensitivity analysis (SA) is to study the variation in the outputs due to a variation in the inputs. Local SA methods typically consider only one input and one output (e.g., one-at-a-time analysis) and are limited to small variations around the nominal operating point. Global sensitivity methods (GSA) focus on output variations over a larger range of inputs, which are considered individually (first order effects) and together (higher order/interaction effects). A classical variance-based GSA method is the method using Sobol's sensitivity indices [21]. The sensitivity indices are obtained via the ANOVA decomposition (Sobol decomposition) of the model outputs. Here, we discuss a brief theory of variance based GSA followed by practical methods to estimate these indices.

2.2.1. Variance-Based Sensitivity Analysis

Consider an d-dimensional input parameter vector $\boldsymbol{\theta}$ with a parameter space Ω_d. The parameters are independent identically distributed (iid) random variables with a probability density function (PDF) $p(\boldsymbol{\theta})$. Since the parameters are iid, $p(\boldsymbol{\theta}) = \prod_i p(\theta_i)$. A physical model $y = f(\boldsymbol{\theta})$ describes the response of a physical system. Note that although y is considered a scalar here, the theory is equally applicable for multi-response systems. The response function can be decomposed into main effects and interactions:

$$f(\boldsymbol{\theta}) := f_0 + \sum_{i=1}^{d} f_i(\theta_i) + \sum_{i=1}^{d-1}\sum_{j>i}^{d} f_{ij}(\theta_i, \theta_j) + \cdots + f_{i_1,\ldots,i_d}(\theta_i, \ldots, \theta_d) \tag{16}$$

The above decomposition is unique, and the following properties are satisfied [21,22]:

$$f_0 = \int_{\Omega_d} f(\boldsymbol{\theta}) p(\boldsymbol{\theta}) d\boldsymbol{\theta} \tag{17}$$

$$\int_{\Omega_{d_k}} f_{i_1,\ldots,i_s} p(\theta_k) d\theta_k = 0, \quad 1 \leq i_1 \leq \cdots < i_s \leq d, k \in i_1, \ldots, i_s \tag{18}$$

From the properties (17) and (18), the terms in the decomposition (16) follow as:

$$\begin{aligned} f_0 &= \mathbb{E}[f(\boldsymbol{\theta})] \\ f_i(\theta_i) &= \mathbb{E}[f(\boldsymbol{\theta})|\theta_i] - f_0 \\ f_{i,j}(\theta_i, \theta_j) &= \mathbb{E}[f(\boldsymbol{\theta})|\theta_i, \theta_j] - f_i(\theta_i) - f_j(\theta_j) - f_0 \\ &\cdots \end{aligned} \tag{19}$$

where $\mathbb{E}[\cdot|\cdot]$ is the conditional expectation. It can be shown that all the summands except f_0 are mutually orthogonal.

Similarly, the variance of the model response can be decomposed as:

$$\mathbb{V}[f(\boldsymbol{\theta})] := D = \sum_{i=1}^{d} D_i + \sum_{1 \leq i < j \leq d} D_{i,j} + \cdots + D_{1,\ldots,d} \tag{20}$$

where $D_{i_1,\cdots,i_s}$ are called the partial variances and are defined as

$$D_{i_1,\ldots,i_s} := \mathbb{V}[f_{i_1,\ldots,i_s}(\theta_{i_1}, \ldots \theta_{i_s})], \quad s = 1, \ldots, M \tag{21}$$

The sensitivity indices for parameter $[\theta_{i_1}, \ldots, \theta_{i_s}]$ are now defined as:

$$S_{i_1,\ldots,i_s} = \frac{D_{i_1,\ldots,i_s}}{D} \tag{22}$$

The first order sensitivity index S_i lies between $[0,1]$, and for purely-additive models, the sum of all first order indices is one.

The total impact of a parameter θ_i is assessed through the total sensitivity index [30]:

$$S_i^T = S_i + \sum_{j<i} S_{j,i} + \sum_{j<k<i} S_{j,k,i} + \cdots + S_{1,\ldots,M} \quad (23)$$

2.2.2. Computation of Sensitivity Indices by Saltelli's Method

In this work, the method proposed by Saltelli et al. [22] was used, which is an extension of the original Monte Carlo sampling-based method of Sobol [21]. For a model with d inputs, the method proceeds as follows:

1. Generate a sample matrix of $N \times 2d$ using the Sobol sequences. The sample matrix is split into two data matrices A (Equation (24)) and B (Equation (25)), each containing half of the samples. N is the number of samples to be used for computing the indices. The order of N can vary between a few hundreds to a few thousands.

$$A = \begin{bmatrix} \theta_1^{(1)} & \theta_2^{(1)} & \cdots & \theta_i^{(1)} & \cdots & \theta_d^{(1)} \\ \theta_1^{(2)} & \theta_2^{(2)} & \cdots & \theta_i^{(2)} & \cdots & \theta_d^{(2)} \\ \cdots & \cdots & \cdots & \cdots & \cdots & \cdots \\ \theta_1^{(N-1)} & \theta_2^{(N-1)} & \cdots & \theta_i^{(N-1)} & \cdots & \theta_d^{(N-1)} \\ \theta_1^{(N)} & \theta_2^{(N)} & \cdots & \theta_i^{(N)} & \cdots & \theta_d^{(N)} \end{bmatrix} \quad (24)$$

$$B = \begin{bmatrix} \theta_{d+1}^{(1)} & \theta_{d+2}^{(1)} & \cdots & \theta_{d+i}^{(1)} & \cdots & \theta_{2d}^{(1)} \\ \theta_{d+1}^{(2)} & \theta_{d+2}^{(2)} & \cdots & \theta_{d+i}^{(2)} & \cdots & \theta_{2d}^{(2)} \\ \cdots & \cdots & \cdots & \cdots & \cdots & \cdots \\ \theta_{d+1}^{(N-1)} & \theta_{d+2}^{(N-1)} & \cdots & \theta_{d+i}^{(N-1)} & \cdots & \theta_{2d}^{(N-1)} \\ \theta_{d+1}^{(N)} & \theta_{d+2}^{(N)} & \cdots & \theta_{d+i}^{(N)} & \cdots & \theta_{2d}^{(N)} \end{bmatrix} \quad (25)$$

2. Define matrix C (Equation (26)) as the matrix with all columns of B except the i^textth column, which is taken from A.

$$C_i = \begin{bmatrix} \theta_{d+1}^{(1)} & \theta_{d+2}^{(1)} & \cdots & \theta_i^{(1)} & \cdots & \theta_2^{(1)} \\ \theta_{d+1}^{(2)} & \theta_{d+2}^{(2)} & \cdots & \theta_i^{(2)} & \cdots & \theta_{2d}^{(2)} \\ \cdots & \cdots & \cdots & \cdots & \cdots & \cdots \\ \theta_{d+1}^{(N-1)} & \theta_{d+2}^{(N-1)} & \cdots & \theta_i^{(N-1)} & \cdots & \theta_{2d}^{(N-1)} \\ \theta_{d+1}^{(N)} & \theta_{d+2}^{(N)} & \cdots & \theta_i^{(N)} & \cdots & \theta_{2d}^{(N)} \end{bmatrix} \quad (26)$$

3. Compute the model output for all the input values in the three matrices A, B, C_i.

$$y_A = f(A) \quad y_B = f(B) \quad y_{C_i} = f(C_i) \quad (27)$$

4. The first order sensitivity indices are estimated as:

$$S_i = \frac{D_i}{D} = \frac{\frac{1}{N}\sum_{j=1}^{N} y_A^{(j)} y_{C_i}^{(j)} - f_0^2}{\frac{1}{N}\sum_{j=1}^{N} (y_A^{(j)})^2 - f_0^2} \quad (28)$$

where,

$$f_0^2 = \frac{1}{N}\left[\sum_{j=1}^{N} y_A^{(j)}\right]^2 \tag{29}$$

The total sensitivity indices are estimated as:

$$S_i^T == 1 - \frac{\frac{1}{N}\sum_{j=1}^{N} y_B^{(j)} y_{C_i}^{(j)} - f_0^2}{\frac{1}{N}\sum_{j=1}^{N}(y_A^{(j)})^2 - f_0^2} \tag{30}$$

The total computational cost of this approach is $N(d+2)$ function evaluations, which is much lower than the N^2 evaluations that would have been required for the brute-force Monte Carlo method.

2.2.3. Computation of Sensitivity Indices Using Arbitrary Polynomial Chaos Expansions

The method described in the previous section requires $N(d+2)$ model evaluations. For models that are computationally expensive, Saltelli's method can quickly become impractical. One way to reduce the computational cost is to approximate the decomposition (16) using meta-modeling techniques. One such approach is the polynomial chaos expansions (PCE). In the PCE approach, the model response is approximated by an infinite sum of orthogonal basis functions. For practical implementation, this series is truncated to M terms.

$$f(\boldsymbol{\theta}) := y_{PCE} = \sum_{i=0}^{M} \alpha_i \Phi_i(\boldsymbol{\theta}) \tag{31}$$

The number of terms M depends on the polynomial order p and the number of random inputs d:

$$M+1 = \frac{(p+d)!}{p!d!} \tag{32}$$

The basis function Φ_i can be formulated using one-dimensional polynomials chosen according to the Wiener–Askey scheme [31]. Using this scheme requires the input parameters to follow one of the five well-defined PDFs (normal, uniform, exponential, beta, and gamma). If any of the inputs do not follow either of these PDFs, they need to be transformed to one. This makes using the Wiener–Askey PCE inefficient. Moreover, in many cases, the input parameter could exist as raw data without an analytical expression of its PDF, making any transformation impossible. In this work, we utilized arbitrary or data-driven polynomial chaos expansions (aPCE). In aPCE, the one-dimensional orthogonal polynomials are constructed through statistical moments of the random inputs and thus do not require them to follow any particular PDF [32]. A one-dimensional polynomial of order p can be generated iff 0 to $2p-1$ order statistical moments exist and are finite. Additionally, to normalize the polynomials, the existence of a finite $2p^{\text{th}}$ moment is necessary [32,33].

Once the basis function is available, the coefficients α_is of the PCE model (31) need to be computed. The methods to compute these coefficient can be categorized as intrusive or non-intrusive methods [34,35]. In this work, the non-intrusive method based on least squares regression was used. This method requires the model to be evaluated at N samples resulting in a vector $\mathbf{y} = [f(\boldsymbol{\theta}^{(0)}), \cdots, f(\boldsymbol{\theta}^{(N)})]$. A set of N linear equations with the M unknown coefficients is then generated as:

$$\underbrace{\begin{bmatrix} f(\boldsymbol{\theta}^{(1)}) \\ f(\boldsymbol{\theta}^{(2)}) \\ \vdots \\ f(\boldsymbol{\theta}^{(N)}) \end{bmatrix}}_{\mathbf{y}} = \underbrace{\begin{bmatrix} \Phi_0(\boldsymbol{\theta}^{(1)}) & \cdots & \Phi_0(\boldsymbol{\theta}^{(1)}) \\ \Phi_0(\boldsymbol{\theta}^{(2)}) & \cdots & \Phi_0(\boldsymbol{\theta}^{(2)}) \\ \vdots & \ddots & \vdots \\ \Phi_M(\boldsymbol{\theta}^{(N)}) & \cdots & \Phi_M(\boldsymbol{\theta}^{(N)}) \end{bmatrix}^{\top}}_{\Lambda^{\top}} \underbrace{\begin{bmatrix} \alpha_0 \\ \alpha_1 \\ \vdots \\ \alpha_M \end{bmatrix}}_{\mathbf{a}} \tag{33}$$

The coefficients can be computed explicitly as:

$$\mathbf{a} = \left(\left(\Lambda\Lambda^{\top} \right)^{-1} \Lambda \right) \mathbf{y} \tag{34}$$

The statistical moments of the model response can be derived based on the PCE coefficients as

$$\mathbb{E}[f(\boldsymbol{\theta})] = \alpha_0 \tag{35}$$

$$D_{PCE} := \mathbb{V}[f(\boldsymbol{\theta})] = \sum_{j=1}^{M} \alpha_j \tag{36}$$

The PCE can be reordered into separate individual and interactive contribution of each parameter, following which, the PCE-based sensitivity indices can be evaluated. This requires defining a set of multiple indices $\mathcal{I}_{k_1,\ldots,k_s}$ as:

$$\mathcal{I}_{k_1,\ldots,k_s} = \left\{ (k_1, \ldots, k_s) : 0 \leq p_k^j \leq p, p_k^j = 0, k \in \{1, \ldots, d\} \setminus \{k_1, \ldots, k_s\} \right\}, \tag{37}$$

where p_k^j is the degree of the univariate polynomial. Using this notation, the sensitivity indices can be expressed as [26,27]:

$$S_i = \frac{\sum_{j \in \mathcal{I}_i} \alpha_j^2}{D_{PCE}} \tag{38}$$

Similarly, higher order sensitivity indices can be obtained as:

$$S_{i_1,\ldots,i_s} = \frac{\sum_{j \in \mathcal{I}_{i_1,\ldots,i_s}} \alpha_j^2}{D_{PCE}} \tag{39}$$

Thus, the computation of PCE-based sensitivity indices requires only N function evaluations rather than $N(2+d)$ evaluations required for Saltelli's method. For PCE, a rough estimate for the minimum number of samples required is $N = 2(M+1)$ [36].

2.3. GSA of the Spray Drying Process

The model was evaluated considering two different spray nozzle configurations: (i) a pneumatic nozzle and (ii) a pressure nozzle. These two configurations were considered to evaluate the impact that the parameters influencing the droplet formation had on the properties of the particulate product. It was also intended to determine if a specific nozzle provided more robustness against any process variability. In a pneumatic nozzle, the gas at high velocity (around sonic) disperses the liquid into droplets. Although many different configurations are available within pneumatic nozzles, a nozzle with parallel flow was considered in this study (Figure 3a). In such nozzles, the droplet size is mainly determined by: (i) the nozzle dimensions, (ii) the liquid properties, (iii) the gas velocity, and (iv) the ratio between gas and liquid flows (ALR). An empirical estimate of the droplet size using these properties is given by Equation (40) [1].

$$d_{32} = Ad_n \left[\frac{We}{\left(1 + \frac{1}{ALR}\right)^2} \right]^B (1 + COh) \tag{40}$$

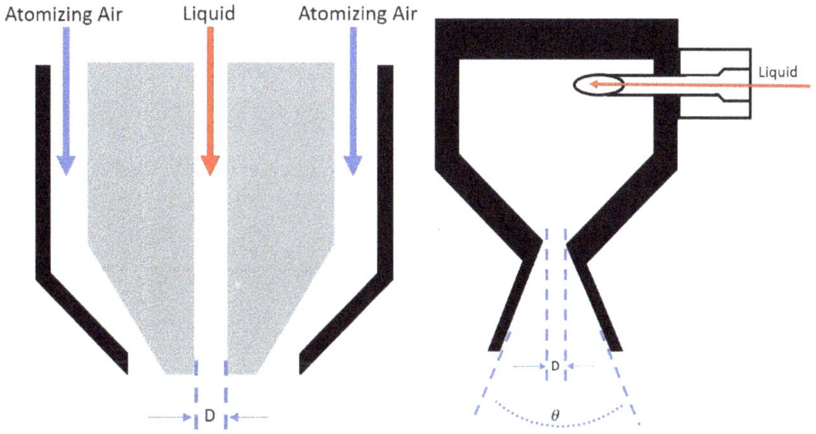

(**a**) Pneumatic nozzle (**b**) Pressure nozzle

Figure 3. Schematics of the two different nozzles of diameter D.

Nozzle diameter and the air liquid ratio enter the above equation explicitly, while the air/liquid properties and gas velocity enters via the Ohnsorge (Oh) and Weber (We) numbers. The three empirical parameters A, B, and C need to be determined through experimental data. In this study, the parameters determined in [1] were used. The dispersion of liquid due to the impact of a gas jet in a pneumatic nozzle occurred when the dynamic pressure of the gas exceeded the internal pressure of the droplets formed. In [37], it was established that the breakup of liquid started in the range $8 < We < 10$. The minimum required gas velocity in the nozzle can be computed from this condition, given the properties of the liquid.

In a pressure nozzle (Figure 3b), the droplets were formed by an abrupt pressure drop at the tip of the nozzle. At this point, the energy of the liquid upstream, in the form of pressure, was converted into velocity. Among the many pressure nozzle configurations available, the swirl atomizer type nozzle was considered in this study. The droplet size in this case was determined by the inner nozzle dimensions, the fluid properties, and the upstream pressure upstream. An empirical relation for this nozzle is given in

Equation (41). This relation relates the droplet size to: (i) the nozzle diameter (d_n), (ii) pressure drop (Δp), (iii) fluid properties via the Reynolds number (Re_p), (iv) and the density ratio [37].

$$\frac{d_{32}}{d_n} = 2.3\Psi^{1/4}\Delta p^{*-1/4}Re_p^{-1/4}\left(\frac{\rho}{\rho_a}\right)^{1/4} \tag{41}$$

The discharge coefficient (Ψ) of the swirl nozzle was calculated using the empirical correlation developed in [38]. The formation of small droplets depends on the formation of a thin sheet of liquid from the tip of the pressure nozzle. According to [37], this condition is achieved with low viscosity liquids for which $Oh < 0.05$. Additionally, a minimum pressure is required to form this hyperbolic sheet in the swirl nozzles.

3. Results

Five main outputs were considered for the sensitivity analysis: (i) the final particle size, (ii) residual solvent content, (iii) particle density, (iv) maximum temperature, and (v) the distance in the chamber at which the equilibrium was reached (if it was reached). The first three outputs can be important CQAs, which need to be monitored. For temperature-sensitive products, the maximum temperature is an important factor to consider, as they might degrade at high temperatures. For the pressure nozzle, the upstream liquid pressure P was included to evaluate the flow energy requirement. The input parameters depending on the nozzle used and their ranges are given in Table 1.

Table 1. Input and output variables considered for the sensitivity analysis with their ranges. Viscosity and upstream liquid pressure are considered only for the pressure nozzle, while ALR is considered only for the pneumatic nozzle.

	Variable	Symbol	Range of Variation	Units
Input variables	Flow of the liquid feed	Q_l	5×10^{-8}–2×10^{-7}	m^3/s
	Flow of the drying gas	$M_{a,d}$	8.5×10^{-4}–8.5×10^{-2}	kg/s
	Initial Temp.of the gas	$T_{a,0}$	350–420	K
	Initial Temp. of the liquid	$T_{l,0}$	300–350	K
	Viscosity of the liquid	ν_l	1×10^{-3}–2×10^{-2}	Pa·s
	Air-to-liquid ratio	ALR	0.5–4	-
Output variables	Particle diameter	d_p	-	μm
	Residual solvent content	W_p	-	-
	Particle density	ρ_p	-	kg/m^3
	Length for equilibrium	Z_{eq}	-	m
	Particle's maximum temperature	$T_{p,max}$	-	K
	Upstream liquid pressure	P	-	bar

All the models were implemented in MATLAB 2017b (The MathWorks Inc., Natick, MA, USA). The system of differential equations was solved using a variable order variable step solver based on numerical difference formulas (*ode15s*), which is known to handle stiff systems well. The sensitivity indices were calculated using Saltelli's approximation and PCE approximation. In the following sections, the two approximations are compared, and then an analysis on the spray drying process is provided.

3.1. Computation of Sensitivity Indices

As analytical calculation of sensitivity indices is almost impossible for complex models, so an approximation based on pseudorandom sampling was used to get a good estimate. Traditionally, sampling-based methods assume that a large sample size (>10,000 samples) is enough to guarantee a valid solution. However, how many samples are enough is a difficult question to answer a priori. In this study, an approach with an iterative increase in the number of samples was used to determine the sample size. Starting from 5000 samples, the sample size was increased by 1000 samples at every iteration. A stopping criterion was defined based on the relative change of the normalized first order and total sensitivity indices. Figure 4 shows the evolution of the mean (μ) and the standard deviation (σ) of the change in normalized sensitivity indices (first order, $S_{ij}/\sum_j S_{ij}$; and total $S_{T,ij}/\sum_j S_{T,ij}$) with the increasing number of samples. The number of samples was considered large enough when the the standard deviation in the relative change was less than 1%.

Based on Figure 4, a set of 23,000 samples was considered large enough for the approximation for a stable estimation of the sensitivity indices. Even though there were smaller sample sizes that met the criterion, it was only after 23,000 samples that the results seemed to stabilize below the defined threshold. With 23,000 samples and five input parameters, the calculation of sensitivity indices required around 161,000 function evaluations!As mentioned before, the current best practices suggest that the sensitivity indices calculated as above can be considered true values [23]. Thus, these will be used to compare the accuracy of sensitivity indices approximated by the PCE approach.

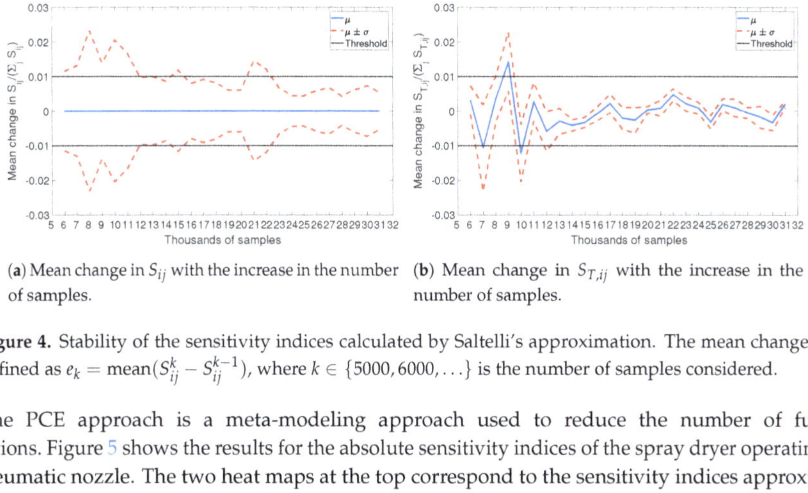

(**a**) Mean change in S_{ij} with the increase in the number of samples.
(**b**) Mean change in $S_{T,ij}$ with the increase in the number of samples.

Figure 4. Stability of the sensitivity indices calculated by Saltelli's approximation. The mean change is defined as $e_k = \text{mean}(S_{ij}^k - S_{ij}^{k-1})$, where $k \in \{5000, 6000, \ldots\}$ is the number of samples considered.

The PCE approach is a meta-modeling approach used to reduce the number of function evaluations. Figure 5 shows the results for the absolute sensitivity indices of the spray dryer operating with the pneumatic nozzle. The two heat maps at the top correspond to the sensitivity indices approximated using Saltelli's method, and those at the bottom were approximated using PCE. Table 2 presents the input ranking based on the total sensitivity indices.

Table 2. Input ranking based on total sensitivity indices computed using Saltelli's approximation (Sal) with 23,000 samples and third order PCE with 200 samples.

Rank	d_p		W_p		ρ_p		Z_{eq}		$T_{p_{max}}$	
	Sal	PCE	Sal	PCE	Sal	PCE	Sal	PCE	Sal	PCE
ALR	1	1	1	1	1	1	2	2	2	2
Q_l	2	2	2	2	4	4	4	4	4	4
$M_{a,d}$	-	-	2	2	2	2	1	1	3	3
$T_{a,0}$	-	-	3	3	3	3	3	3	1	1
$T_{l,0}$	-	-	-	-	-	-	-	-	4	3

The rank was assigned assuming that a difference smaller than 5% between the sensitivity indices was insignificant. Consequently, any sensitivity index below 0.05 was considered equal to zero and the output assumed to be independent of the input. Similarly, the same rank was given to multiple inputs whose sensitivity indices did not differ significantly.

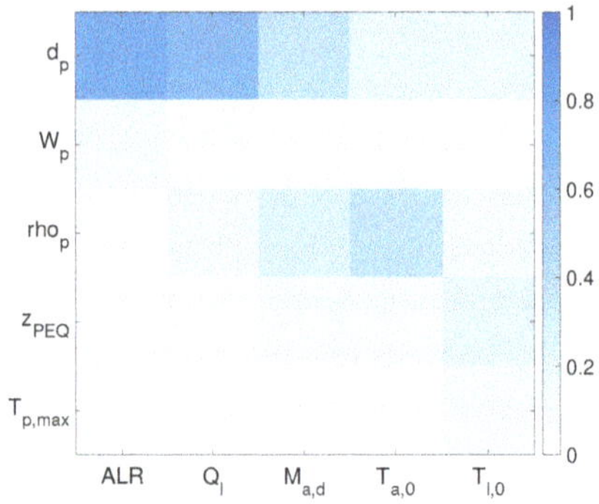

(**a**) First order sensitivity indices S_i (Saltelli).

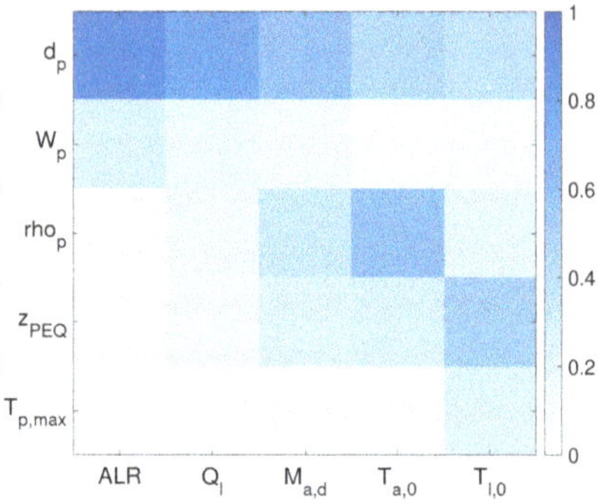

(**b**) Total sensitivity indices $S_{T,i}$ (Saltelli).

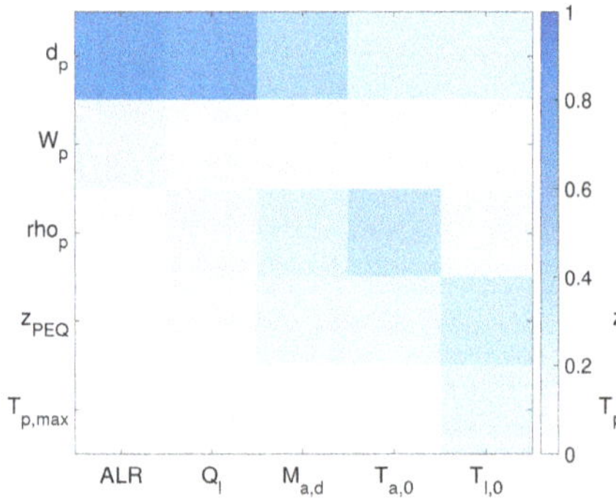

(**c**) First order sensitivity indices S_i (PCE).

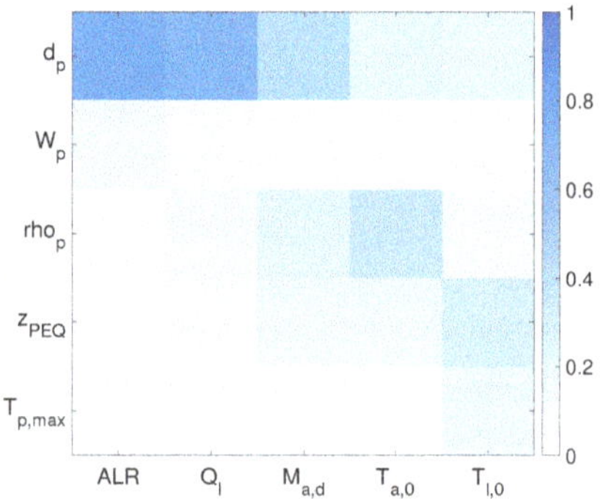

(**d**) Total sensitivity indices $S_{T,i}$ (PCE).

Figure 5. Qualitative comparison of sensitivity heat maps for a spray dryer with the pneumatic nozzle computed using Saltelli's approximation with 23,000 samples and third order PCE with 200 samples. The x-axis shows the operating parameters, and the y-axis shows the output parameters considered.

The input ranking obtained by both the Saltelli and PCE approach was the same, except for the output $T_{p,max}$. In this case, although the first order sensitivity indices were comparable, total sensitivity indices showed significant variation from the ones calculated by Saltelli's method. A plausible reason for this could be that $T_{p,max}$ was extremely nonlinear in inputs, and the correct estimation would require a higher order polynomial. This is evident from Table 3. With increasing order, nonlinearity was captured with much more ease. However, the number of samples required to evaluate the sensitivity indices also increased with the order of the PCE. The benefits on computational time provided by the PCE approach outweighed the slight decrease in accuracy. Even if a high order PCE was used, the number of function evaluation was at least an order of magnitude less than the number of function evaluations for Saltelli's approach.

Table 3. Change in total sensitivity indices ($S_{T,i}$) for $T_{p,max}$ with increasing polynomial order (p) in the PCE approach.

Parameter	PCE ($p = 3$)	PCE ($p = 4$)	PCE ($p = 5$)	PCE ($p = 6$)	PCE ($p = 7$)	Saltelli
ALR	0.3826	0.4481	0.4444	0.5030	0.5031	0.4920
Q_l	0.0706	0.1652	0.1789	0.2975	0.2352	0.2350
$M_{a,d}$	0.1596	0.2315	0.3127	0.3825	0.3498	0.3582
$T_{a,0}$	0.4883	0.4626	0.4463	0.5011	0.4728	0.5352
$T_{l,0}$	0.2341	0.2514	0.2939	0.1907	0.2964	0.2289
# of Samples	200	400	600	1000	2000	23,000
# of Function Evaluations	200	400	600	1000	2000	161,000

3.2. Global Sensitivity Analysis of Spray Dryer with a Pneumatic Nozzle

Based on the previous discussion and to avoid excessive computational costs, the PCE approach was used for the sensitivity analysis of the two spray dryers. A fourth order PCE was used to determine the sensitivity indices, and 600 samples were used to compute the coefficients of the expansion.

Figure 6 depicts the sensitivity indices for a spray dryer operating with the pneumatic nozzle.

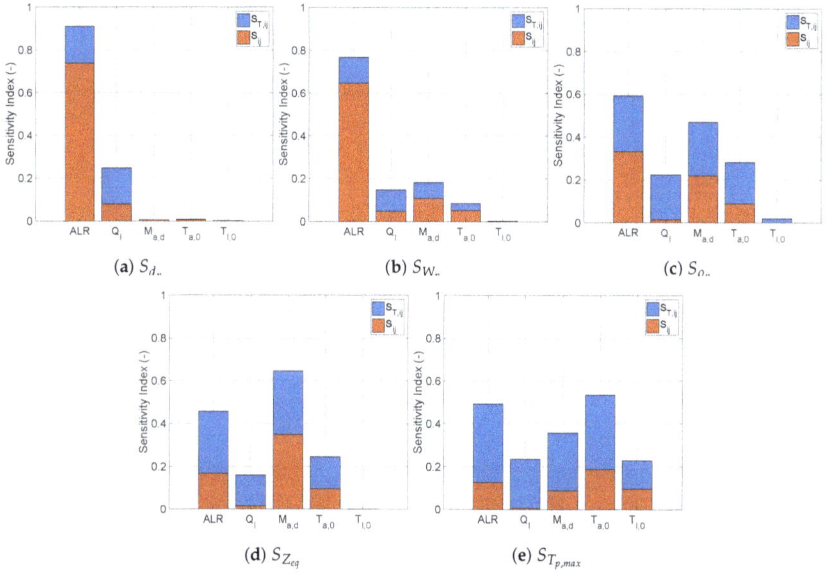

Figure 6. Sensitivity indices ($S_i, S_{T,i}$ $\forall i \in \{d_p, W_p, \rho_p, Z_{eq}, T_{p,max}\}$) for the spray dryer operating with a pneumatic nozzle calculated using fourth order PCE with 600 samples.

Based on the response to the inputs depicted in Figure 6, the output variables could be divided into two groups. The first group consisted of the variables whose responses were mostly determined by the individual change of a single input. The other group contained outputs whose response involved significant high order input interactions. In the first group, the particle diameter and the residual humidity were characterized by a strong first order effect from the ALR along with a minor influence from the other inputs. For the particle size, only the liquid flow rate had a significant effect apart from the ALR. For the residual solvent content, the flow rate and temperature of the drying air played a minor role. This means that in a spray dryer equipped with a pneumatic nozzle, the final particle size and residual humidity were conditioned by the droplet formation, which was determined by the amount of atomization gas used with the liquid flow rate.

In Figure 7, the scatter plots illustrate the solutions for d_p and W_p, from which samples were taken to perform the sensitivity calculations. The blue dots correspond to a product that has reached its drying equilibrium, while red dots correspond to non-equilibrium. It can be noticed that most of the variation occurred at non-equilibrium conditions. It can also be seen that the ALR determined whether equilibrium was reached, and equilibrium was reached only at high ALR. The low spread in the particle size data showed that the particle size was mostly dominated by the ALR. With the increase in the ALR, the spread of the final particle size reduced, and the possible solutions tended to concentrate close to the equilibrium solution. This is the result of the small droplet size obtained from the pneumatic nozzle at high ALR. Since the droplet size was already very close to the equilibrium size of the particle, other conditions in the process did not have a major impact, and the spread of the possible end particle size was very narrow. Differently, in the case of the residual solvent W_p, its large spread of solutions indicated that other inputs remained influential at any value of ALR. For high ALR, the final humidity of the particle at a non-equilibrium condition would depend also on the number of droplets in the chamber; the larger the number of droplets, the more solvent needed to evaporate to reduce their humidity. Thus, for the pneumatic nozzle, the particle size could be tuned solely by the ALR, whereas an accurate control over the residual solvent content required tuning the liquid feed rate as well.

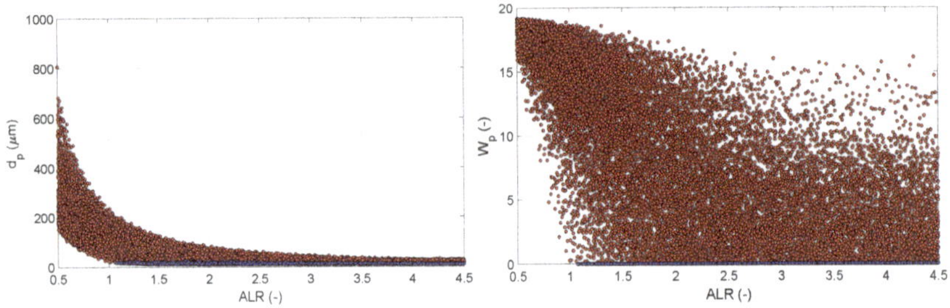

(**a**) ALR vs d_p (**b**) ALR vs W_p

Figure 7. Scatter plots of final particle size d_p (**a**) and residual moisture content W_p with respect to ALR (**b**). The red dots indicate the values at non-equilibrium conditions, while blue dots represent the values at equilibrium. The figures clearly indicate the significant effect ALR has on particle size, along with its effect on residual solvent content. This effect is, however, limited only to the non-equilibrium zone.

For the second group of outputs (ρ_p, Z_{eq}, and $T_{p,max}$), the ALR was not the sole significant input. The contributions of other inputs varied with every output, but a similarity could be observed between the responses of ρ_p and Z_{eq}. As could be expected, the input temperature played a greater role in the determining the maximum particle temperature. The liquid feed temperature was relevant only for $T_{p,max}$. Figure 8 presents the scatter plots of the length required to reach equilibrium with respect to ALR and $M_{a,d}$. The scatter plots confirmed that ALR had a significant influence on the point at which the equilibrium was reached. For ALR below one, all solutions were in the non-equilibrium region. Once this threshold value of ALR was crossed, the scatter suggested that the point at which the equilibrium was reached was determined by more factors than just the ALR. The scatter plot for the relation between Z_{eq} and $M_{a,d}$ showed that an increase in the amount of drying air tended to move the equilibrium point towards the bottom of the dryer.

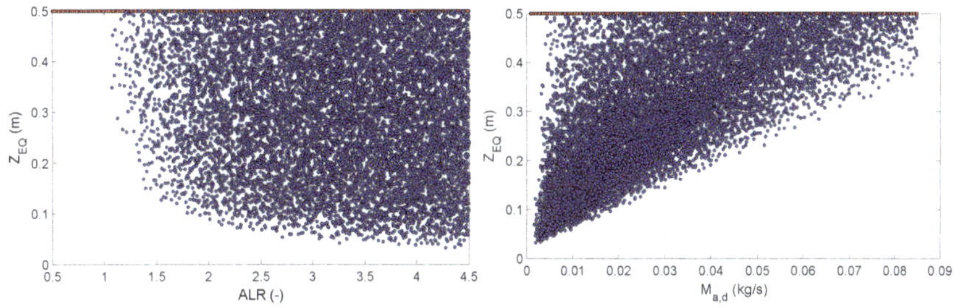

Figure 8. Scatter plots of equilibrium length Z_{eq} with respect to ALR (**a**) and the mass flow rate of the drying air $M_{a,d}$ (**b**). The red dots indicate the values at non-equilibrium conditions, while blue dots represent the values at equilibrium.

3.3. Global Sensitivity Analysis of the Spray Dryer with a Pressure Nozzle

Figure 9 depicts the sensitivity indices for the spray dryer operating with a pressure nozzle.

Analogously to the pneumatic nozzle, the outputs could be divided into two groups. The first group consisted of the particle diameter and the upstream liquid pressure, which were dominated by first order effects corresponding to the droplet formation. For liquid feed pressure, this was expected, as the pressure required for the liquid to go through the nozzle was independent of the drying. Contrary to the pneumatic nozzle, the residual solvent content was not part of this group. Since there was no additional process parameter affecting the droplet formation, the feed flow rate Q_l became the most influential variable for the end particle size.

Along with Q_l, the viscosity of the liquid also had a significant effect on the particle diameter. In practice, the variation in the feed flow might be intentional or induced by uncontrolled variability in the nozzle characteristics when the spray dryer is operated at a fixed upstream pressure. The variability of the viscosity might be uncontrolled due to the properties of the material added to the feed mixture. Figure 10 depicts the scatter of d_p with respect to Q_l and ν_l. The dependence of the particle size on these two variables is clearly visible from these figures. Additionally, these relations were also maintained in equilibrium conditions. However, at equilibrium, the variation was reduced. The narrow spread of the d_p scatter for Q_l explained the significance of Q_l.

The residual solvent content showed a higher level of interaction between the droplet formation and the drying process. The contributions were distributed along the evaluated process conditions. Along with the feed flow rate and viscosity, the flow rate of drying air and its initial temperature also had a significant impact on the residual solvent content.

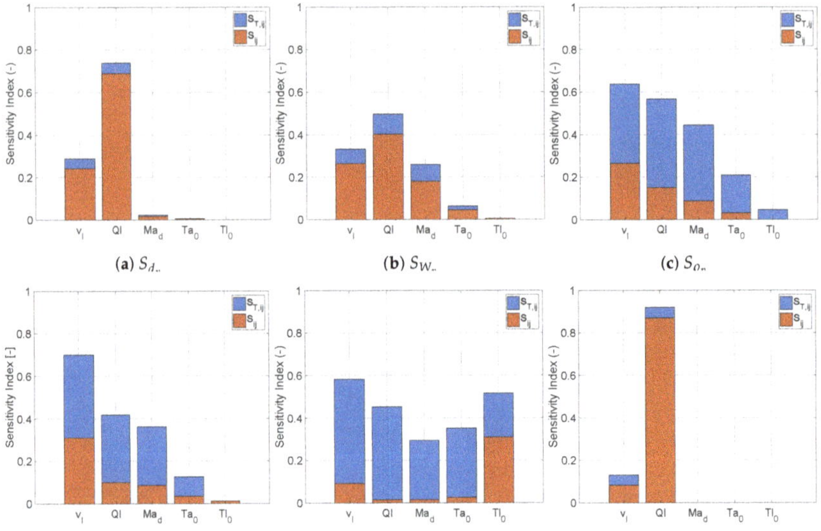

(a) S_{d_p} (b) S_{W_p} (c) S_{ρ_p}

(d) $S_{Z_{eq}}$ (e) $S_{T_{p,max}}$ (f) S_P

Figure 9. Sensitivity indices ($S_i, S_{T,i}$ $\forall i \in \{d_p, W_p\rho_p Z_{eq} T_{p,max} P\}$) for the spray dryer operating with a pressure nozzle calculated using fourth order PCE with 600 samples.

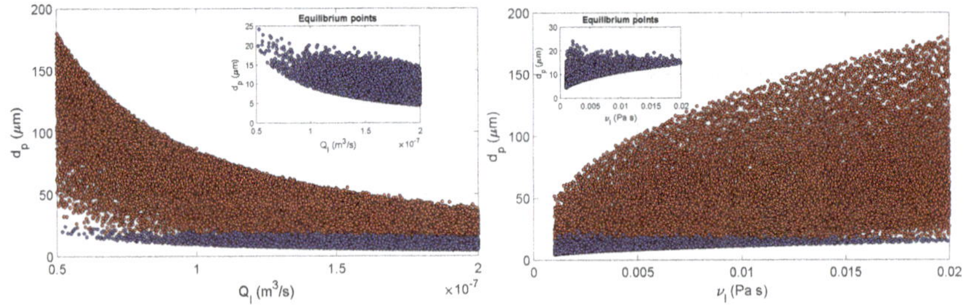

(a) Q_l vs d_p (b) ν_l vs d_p

Figure 10. Scatter plots of final particle size d_p with respect to feed flow rate Q_l (**a**) and feed viscosity ν_l (**b**). The red dots indicate the values at non-equilibrium conditions, while blue dots represent the values at equilibrium. The equilibrium points are also shown explicitly in the insets. Unlike the pneumatic nozzle, the pressure nozzle conditions also influence the equilibrium values.

The second group consisted of particle density, the distance at which equilibrium was reached, and the maximum temperature reached. As was the case with the pneumatic nozzle, the sensitivity profile of ρ_p and Z_{eq} for the pressure nozzle was also similar. These two outputs showed high sensitivity to the viscosity, the feed flow rate, and the flow rate of drying gas. Figure 11 shows the scatter of Z_{eq} with respect to ν_l and $M_{a,d}$. The former can easily be identified as the most influential parameter. In Figure 11, the scatter cloud was made primarily of blue dots, i.e., results achieving equilibrium at different lengths inside the drying chamber, and all other results at a non-equilibrium condition had a value $Z_{eq} = 0.5$ m. This shows that for product leaving the dryer at equilibrium, the viscosity had a major impact on the particle size, and in turn also on the length required to reach the equilibrium. Figure 11 illustrates how the same drying parameter might affect in a different way an output variable due to the difference in the interactions with the parameters from the two different nozzles. The direct comparison of this plot with its equivalent in Figure 8 shows that $M_{a,d}$ had a much more significant individual impact on Z_{eq} for the pneumatic nozzle. In case of the pressure nozzle, the velocity of the droplets in the dryer chamber was influenced both by the pressure drop (P) in the nozzle and the $M_{a,d}$ determining the velocity of the gas in the chamber, leading to a much more entropic scatter.

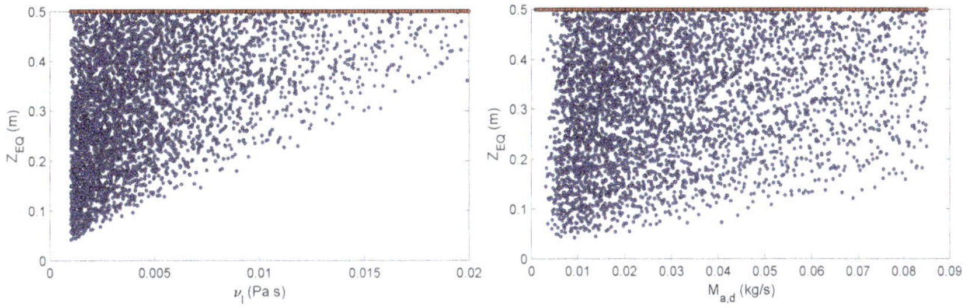

(**a**) nu_l vs Z_{eq} (**b**) $M_{a,d}$ vs Z_{eq}

Figure 11. Scatter plots of equilibrium length Z_{eq} with respect to the feed viscosity ν_l (**a**) and mass flow rate of the drying air $M_{a,d}$ (**b**). The red dots indicate the values at non-equilibrium conditions, while blue dots represent the values at equilibrium.

4. Conclusions

Based on the discussion above, it can be concluded that PCE was an attractive approach to reduce the computational burden required for a global sensitivity analysis. The quantitative values of the sensitivity indices computed using PCE approach, although different, were still comparable to Saltelli's approximation. The input ranking obtained, which was arguably more important than the quantitative values themselves, was the same if a proper PCE order was used. It has to be noted that if the process were highly nonlinear, a low order PCE could lead to erroneous results. However, there is no a priori way of determining the nonlinearity associated with any output. In such situations, an iterative approach with increasing the order of PCE is recommended. Even with such an iterative approach, the number of function evaluations required for the PCE approach were orders of magnitude fewer than the quasi Monte Carlo approach. For the case study considered above, use of a fourth order PCE instead of a stable quasi Monte Carlo approach led to around a 99.6% reduction in computational cost. Even if a seventh order PCE was used, the computational savings were around 96%.

The GSA of the spray drying model allowed the system to be characterized based on the dominant relationships between the CQAs and the CPPs. This analysis provides a detailed view on how the phenomena occurring during atomization and drying interact and determine the output response. Based on this study, it was possible to discriminate the behavior of the spray dryer depending on the type of nozzle.

The sensitivity patterns demonstrated that independent of the nozzle used, mean particle size was affected predominantly by the atomization phenomena, while others' outputs were affected by an interaction between the atomization and drying parameters. The type of nozzle to be used will always depend on the application. Given the presence of the ALR, the pneumatic nozzle had more flexibility for the control of the final particle size. However, this also meant that a tight ALR control was required as a small variance on this parameter would have a large impact on the particle size. For the pressure nozzle, a strong dependency was established along all outputs on the feed flow rate, which additionally will determine the throughput of the unit. This dependency was strengthened by the fact that both the initial droplet size and velocity were extremely affected by changes in the liquid flow rate. This diminished the effect of the drying parameters. This was specifically clear for particle density and the distance to reach equilibrium. The direct effect of the drying flow rate for these two outputs was found to be less important in the case of the pressure nozzle compared to the pneumatic nozzle.

Future work will be focused on applying other computational intelligence methods, e.g., fuzzy logic and genetic programming, to compare their performance for GSA, but also for knowledge discovery purposes.

Author Contributions: Conceptualization: S.B., C.A.M.L., and J.V.I.; methodology: S.B., C.A.M.L., and J.V.I.; software: S.B. and C.A.M.L.; validation: C.A.M.L.; formal analysis S.B. and C.A.M.L.; investigation: S.B. and C.A.M.L.; resources: S.B., C.A.M.L., and J.V.I.; writing, original draft preparation: S.B. and C.A.M.L.; writing, review and editing: S.B., C.A.M.L., and J.V.I.; visualization: S.B. and C.A.M.L.; supervision: J.V.I.; project administration: J.V.I.; funding acquisition: J.V.I.

Funding: This work was supported by Fund for Scientific Research - Flanders project FWO-G0863.18. S.B. holds an IWT (VLAIO) Baekeland Grant (IWT150715). C.A.M.L. holds a VLAIO Baekeland Grant (HBC2017.0239).

Conflicts of Interest: The authors declare no conflict of interest.

Abbreviations

The following abbreviations are used in this manuscript:

ALR		Air liquid ratio
CMA		Critical material attributes
CPP		Critical process parameters
CQA		Critical quality attributes
GSA		Global sensitivity analysis
PCE		Polynomial chaos expansions
QbD		Quality by design
List of Symbols		
α	W/m^2K	Heat transfer coefficient
β	kg/m^2s	Mass transfer coefficient
ν	Pa s	Viscosity
ρ	kg/m^3	Density
Ψ	-	Discharge coefficient
C_D	-	Drag coefficient
d	m	Diameter
$\mathcal{D}_{eff}$	m^2/s	Effective diffusivity
g	m/s^2	Acceleration due to gravity
ΔH	j/kg	Enthalpy of evaporation
m_s	kg	Solid mass
$\dot{m}_v$	kg/m^2s	Mass transfer rate
M	kg/s	Mass flow rate
$\tilde{M}$	kg/mol	Molecular weight
Nu	-	Nusselt number
Oh	-	Ohnsorge number

P	Pa	Pressure
P_v	Pa	Vapor pressure
Q	m^3/s	Volumetric flow rate
Re	-	Reynolds number
S_i	-	First order sensitivity index
$S_{T,i}$	-	Total sensitivity index
T	K	Temperature
U	W/m^2K	Overall heat transfer coefficient
v	m/s	Velocity
W	-	Residual solvent content
We	-	Weber number
Y_b	-	Gas moisture content
Y_{sat}	-	Gas saturation moisture content
Z	m	Axial distance in the spray drying chamber
Subscripts		
a		Air/gas
c		Critical
eq		Equilibrium
l		Feed liquid
n		Nozzle
p		Particle
s		Solids
w		Water/solvent

References

1. Cotabarren, I.M.; Bertín, D.; Razuc, M.; Rairez-Rigo, M.V.; Pina, J. Modelling of the spray drying process for particle design. *Chem. Eng. Res. Des.* **2018**, *132*, 1091–1104. [CrossRef]
2. Schuck, P.; Jeantet, R.; Bhandari, B.; Chen, X.D.; Perrone, Í.T.; de Carvalho, A.F.; Fenelon, M.; Kelly, P. Recent advances in spray drying relevant to the dairy industry: A comprehensive critical review. *Dry. Technol.* **2016**, *34*, 1773–1790. [CrossRef]
3. Shishir, M.R.I.; Chen, W. Trends of spray drying: A critical review on drying of fruit and vegetable juices. *Trends Food Sci. Technol.* **2017**, *65*, 49–67. [CrossRef]
4. Poozesh, S.; Bilgili, E. Scale-up of pharmaceutical spray drying using scale-up rules: A review. *Int. J. Pharm.* **2019**, *562*, 271–292. [CrossRef] [PubMed]
5. Sosnik, A.; Seremeta, K.P. Advantages and challenges of the spray-drying technology for the production of pure drug particles and drug-loaded polymeric carriers. *Adv. Colloid Interface Sci.* **2015**, *223*, 40–54. [CrossRef] [PubMed]
6. Fatnassi, M.; Tourné-Péteilh, C.; Peralta, P.; Cacciaguerra, T.; Dieudonné, P.; Devoisselle, J.M.; Alonso, B. Encapsulation of complementary model drugs in spray-dried nanostructured materials. *J. Sol-Gel Sci. Technol.* **2013**, *68*, 307–316. [CrossRef]
7. Cheow, W.S.; Li, S.; Hadinoto, K. Spray drying formulation of hollow spherical aggregates of silica nanoparticles by experimental design. *Chem. Eng. Res. Des.* **2010**, *88*, 673–685. [CrossRef]
8. Gharsallaoui, A.; Roudaut, G.; Chambin, O.; Voilley, A.; Saurel, R. Applications of spray-drying in microencapsulation of food ingredients: An overview. *Food Res. Int.* **2007**, *40*, 1107–1121. [CrossRef]
9. Vincente, J.; Pinto, J.; Menezes, J.; Gaspar, F. Fundamental analysis of particle formation in spray drying. *Powder Technol.* **2013**, *247*, 1–7. [CrossRef]
10. Nandiyanto, A.B.D.; Okuyama, K. Progress in developing spray-drying methods for the production of controlled morphology particles: From the nanometer to submicrometer size ranges. *Adv. Powder Technol.* **2011**, *22*, 1–19. [CrossRef]

11. Debevec, V.; Srčič, S.; Horvat, M. Scientific, statistical, practical, and regulatory considerations in design space development. *Drug Dev. Ind. Pharm.* **2018**, *44*, 349–364. [CrossRef] [PubMed]
12. Petersen, L.N.; Poulsen, N.K.; Niemann, H.H.; Utzen, C.; Jørgensen, J.B. An experimentally validated simulation model for a four-stage spray dryer. *J. Process Control* **2017**, *57*, 50–65. [CrossRef]
13. Ferrari, A.; Gutiérrez, S.; Sin, G. Modeling a production scale milk drying process: Parameter estimation, uncertainty and sensitivity analysis. *Chem. Eng. Sci.* **2016**, *152*, 301–310. [CrossRef]
14. Zhang, X.; Pei, Y.; Xie, D.; Chen, H. Modeling Spray Drying of Redispersible Polyacrylate Powder. *Dry. Technol.* **2014**, *32*, 222–235. [CrossRef]
15. Mezhericher, M.; Levy, A.; Borde, I. Multi-Scale Multiphase Modeling of Transport Phenomena in Spray-Drying Processes. *Dry. Technol.* **2015**, *33*, 2–23. [CrossRef]
16. Juaber, H.; Afshar, S.; Xiao, J.; Chen, X.D.; Selomulya, C.; Woo, M.W. On the importance of droplet shrinkage in CFD-modeling of spray drying. *Dry. Technol.* **2018**, *36*, 1785–1801. [CrossRef]
17. Baldinger, A.; Clerdent, L.; Rantanen, J.; Yang, M.; Grohganz, H. Quality by design approach in the optimization of the spray-drying process. *Pharm. Dev. Technol.* **2012**, *17*, 389–397. [CrossRef]
18. Lebrun, P.; Krier, F.; Mantanus, J.; Grohganz, H.; Yang, M.; Rozet, E.; Boulanger, B.; Evrard, B.; Rantanen, J.; Hubert, P. Design space approach in the optimization of the spray-drying process. *Eur. J. Pharm. Biopharm.* **2012**, *80*, 226–234. [CrossRef]
19. Razuc, M.; Piña, J.; Ramírez-Rigo, M.V. Optimization of Ciprofloxacin Hydrochloride Spray-Dried Microparticles for Pulmonary Delivery Using Design of Experiments. *AAPS PharmSciTech* **2018**, *19*, 3085–3096. [CrossRef]
20. Ingvarsson, P.T.; Yang, M.; Mulvad, H.; Nielsen, H.M.; Rantanen, J.; Foged, C. Engineering of an inhalable dda/tdb liposomal adjuvant: A quality-by-design approach towards optimization of the spray drying process. *Pharm. Res.* **2013**, *30*, 2772–2784. [CrossRef]
21. Sobol, I. Global sensitivity indices for nonlunear mathematical modesl and their Monte Carlo estimates. *Math. Comput. Simul.* **2001**, *55*, 271–280. [CrossRef]
22. Saltelli, A.; Ratto, M.; Andres, T.; Campolongo, F.; Cariboni, J.; Gatelli, D.; Saisana, M.; Tarantola, S. *Global Sensitivity Analysis: The Primer*; John Wiley & Sons Ltd.: Chichester, UK, 2008.
23. Saltelli, A.; Annoni, P.; Azzini, I.; Campolongo, F.; Ratto, M.; Tarantola, S. Variance based sensitivity analysis of model output. Design and estimator for the total sensitivity index. *Comput. Phys. Commun.* **2010**, *181*, 259–270. [CrossRef]
24. Todri, E.; Amenaghawon, A.; Jimenez del Val, I.; Leak, D.; Kontoravdi, C.; Kucherenko, S.; Shah, N. Global sensitivity analysis and meta-modeling of an ethanol production process. *Chem. Eng. Sci.* **2014**, *114*, 114–127. [CrossRef]
25. Crestaux, T.; Le Maître, O.; Martinez, J.M. Polynomial chaos expansion for sensitivity analysis. *Reliab. Eng. Syst. Saf.* **2009**, *94*, 1161–1172. [CrossRef]
26. Blatman, G.; Sudret, B. Efficient computation of global sensitivity indices using sparse polynomial chaos expansions. *Reliab. Eng. Syst. Saf.* **2010**, *95*, 1216–1229. [CrossRef]
27. Sudret, B. Global sensitivity analysis using polynomial chaos expansions. *Reliab. Eng. Syst. Saf.* **2008**, *93*, 964–979. [CrossRef]
28. Negiz, A.; Lagergren, E.S.; Cinar, A. Mathematical models of cocurrent spray drying. *Ind. Eng. Chem. Res.* **1995**, *34*, 3289–3302. [CrossRef]
29. Ranz, W.; Marshall, W. Evaporation from drops. Parts I and II. *Chem. Eng. Prog.* **1952**, *48*, 141–146, 173–180.
30. Homma, T.; Saltelli, A. Importance measures in global sensitivity analysis of nonlinear models. *Reliab. Eng. Syst. Saf.* **1996**, *52*, 1–17. [CrossRef]
31. Xiu, D.; Karniadakis, G. The Wiener–Askey polynomial chaos for stochastic differential equations. *SIAM J. Sci. Comput.* **2002**, *24*, 619–644. [CrossRef]
32. Oladyshkin, S.; Nowak, W. Data-driven uncertainty quantification using the arbitrary polynomial chaos expansion. *Reliab. Eng. Syst. Saf.* **2012**, *106*, 179–190. [CrossRef]
33. Yang, S.; Xiong, F.; Wang, F. Polynomial Chaos Expansion for Probabilistic Uncertainty Propagation. In *Uncertainty Quantification and Model Calibration*; Hessling, J.P., Ed.; IntechOpen: Rijeka, Croatia, 2017; Chapter 2.

34. Bhonsale, S.; Nimmegeers, P.; Telen, D.; Paulson, J.A.; Mesbah, A.; Van Impe, J. On the implementation of generalized polynomial chaos in dynamic optimization under stochastic uncertainty: A user perspective. In *Computer Aided Chemical Engineering*; Kiss, A.A., Zondervan, E., Lakerveld, R., Özkan, L., Eds.; Elsvier: Rijeka, Croatia, 2017; Chapter 1.
35. Nimmegeers, P.; Telen, D.; Logist, F.; Van Impe, J. Dynamic optimization of biological networks under parametric uncertainty. *BMC Syst. Biol.* **2016**, *10*, 86. [CrossRef] [PubMed]
36. Bhonsale, S.; Telen, D.; Stokbroekx, B.; Van Impe, J. An Analysis of Uncertainty Propagation Methods Applied to Breakage Population Balance. *Processes* **2018**, *6*, 255. [CrossRef]
37. Walzel, P. Spraying and Atomization of liquids. In *Ullmann's Encyclopedia of Industrial Chemestry*; Wiley-VCH Verlag GmbH: Weinheim, Germany, 2012; pp. 79–98.
38. Wimmer, E.; Brenn, G. Viscous effects on flows through pressure-swirl atomizers. In Proceedings of the 12th Triennial International Conference on Liquid Atomization and Spray Systems (ICLASS 2012), Heidelberg, Germany, 2–6 September 2012.

processes

MDPI

Article

Explicit Residence Time Distribution of a Generalised Cascade of Continuous Stirred Tank Reactors for a Description of Short Recirculation Time (Bypassing)

Peter Toson [1,*], Pankaj Doshi [2] and Dalibor Jajcevic [1]

[1] Research Center Pharmaceutical Engineering GmbH, Inffeldgasse 13, 8010 Graz, Austria
[2] Worldwide Research and Development, Pfizer Inc. Groton, CT 06340, USA
* Correspondence: peter.toson@rcpe.at; Tel.: +43-316-873-30980

Received: 30 July 2019; Accepted: 5 September 2019; Published: 10 September 2019

Abstract: The tanks-in-series model (TIS) is a popular model to describe the residence time distribution (RTD) of non-ideal continuously stirred tank reactors (CSTRs) with limited back-mixing. In this work, the TIS model was generalised to a cascade of n CSTRs with non-integer non-negative n. The resulting model describes non-ideal back-mixing with $n > 1$. However, the most interesting feature of the n-CSTR model is the ability to describe short recirculation times (bypassing) with $n < 1$ without the need of complex reactor networks. The n-CSTR model is the only model that connects the three fundamental RTDs occurring in reactor modelling by variation of a single shape parameter n: The unit impulse at $n \rightarrow 0$, the exponential RTD of an ideal CSTR at $n = 1$, and the delayed impulse of an ideal plug flow reactor at $n \rightarrow \infty$. The n-CSTR model can be used as a stand-alone model or as part of a reactor network. The bypassing material fraction for the regime $n < 1$ was analysed. Finally, a Fourier analysis of the n-CSTR was performed to predict the ability of a unit operation to filter out upstream fluctuations and to model the response to upstream set point changes.

Keywords: residence time distribution; continuous stirred tank reactor; bypassing; Fourier analysis; continuous manufacturing

1. Introduction

The pharmaceutical industry is currently transforming batch production processes to continuous manufacturing. Continuous manufacturing offers several technical and economic advantages compared to batch processes, such as lower downtimes, better process control, smaller footprints, and ease of scale-up by extending time [1–4]. Better process understanding and control lead ultimately to improved quality of the final product, in a quality-by-design (QdB) framework [3,5,6]. However, material tracking from raw material to finished product remains challenging.

The tool of choice for modelling the flow of material in a continuous process is the prediction and modelling of residence time distributions (RTDs). Each unit operation (e.g., blending, granulation, tableting) is characterised by its RTD. The individual RTDs are then chained together by convolution integrals, in order to calculate the RTD of the overall process. With this information it is possible to predict how long the material remains, on average, in the process (mean residence time, MRT), the response of the system to fluctuations in the material stream (e.g., feeder refills), and to develop process control strategies [7–11].

Residence time distribution modelling is not only used to describe a complete continuous manufacturing line; it is also utilised to describe the complex behaviour of single unit operations within reactor networks. A reactor network contains multiple connected ideal or non-ideal reactors, with a known analytical RTD. The two most common types are the continuous stirred tank reactor (CSTR) and the plug flow reactor (PFR). These reactor types are too idealised to correctly model the behaviour of a real reactor. However, if these basic models are combined in reactor networks, it is possible to describe the behaviour of a real unit operation, including effects such as dead zones, non-ideal back mixing, and/or bypassing effects. This approach is not limited to pharmaceutical processes, but is also utilised, for example, in chemical reaction engineering and modelling of water treatment processes [12–14].

A well-established model for non-ideal CSTRs is the tanks-in-series (TIS) model, describing a cascade of n CSTRs. This model has an analytical solution for integer n, that was first published by MacMullin and Weber [15] and is now part of standard chemical engineering literature [16,17]. A long chain of tanks-in-series results into a sharp RTD peak around the MRT, which is better described by the diffusion model. The diffusion model accounts for axial diffusion in a non-ideal PFR and produces broader peaks in the RTD with increasing diffusion [18–20]. If the fluid velocities inside the PFR are high, compared to the total length of the reactor but still in the laminar flow regime, the RTD is dominated by the parabolic shape of the velocity profile and not by slow diffusion processes. In this case, the RTD is better explained by the convection model [21–24].

Martin [25] generalised the TIS model to non-integer $n \geq 1$ to fine-tune the resulting RTD. This generalisation has been used as a stand-alone model and as a building block in reactor networks [13,24,26–28]. Toson et al. [29] used the model with a narrow parameter range $0.5 < n < 1.1$ to fit the RTD of a continuous powder mixer and linked n to the quality of the mixing process.

The n-CSTR model discussed in this work is an extension of Martin's model [25]. Just like Martin's model, a non-integer value of $n > 1$ allows fine tuning of the RTD shape for varying degrees of limited back-mixing. The n-CSTR model extends the value range to $n < 1$, in order to model bypassing conditions. The n-CSTR model has only one shape parameter, n, that can be varied to reach the unit impulse at $n \to 0$, the ideal CSTR at $n = 1$, and the ideal PFR at $n \to \infty$, with the same analytical form. The tanks-in-series model, the diffusion model, and the convection model only connect two of these fundamental RTD shapes (see Figures 1 and 2)

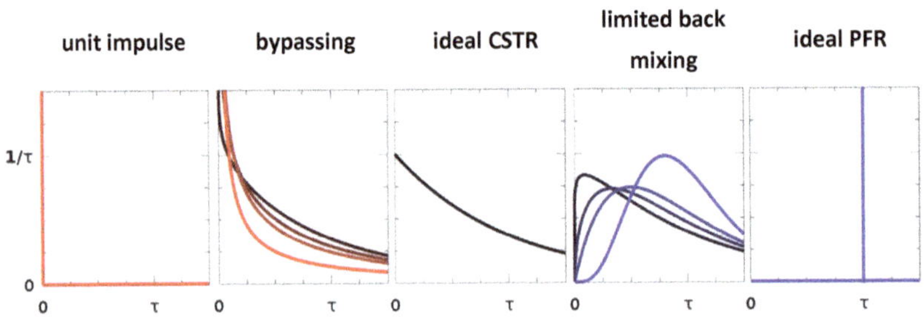

	unit impulse	bypassing	ideal CSTR	limited back mixing	ideal PFR
TIS equation (8)	n/a	n/a	$n = 1$	$n = 2,3,4 \ldots$	$n \to \infty$
Diffusion equation (10)	n/a	n/a	$D \to \infty$ $\mathrm{Bo} = 0$	$D > 0$ $\mathrm{Bo} > 0$	$D = 0$ $\mathrm{Bo} \to \infty$
n-CSTR equation (16)	$n = 0$	$0 < n < 1$	$n = 1$	$n > 1$	$n \to \infty$

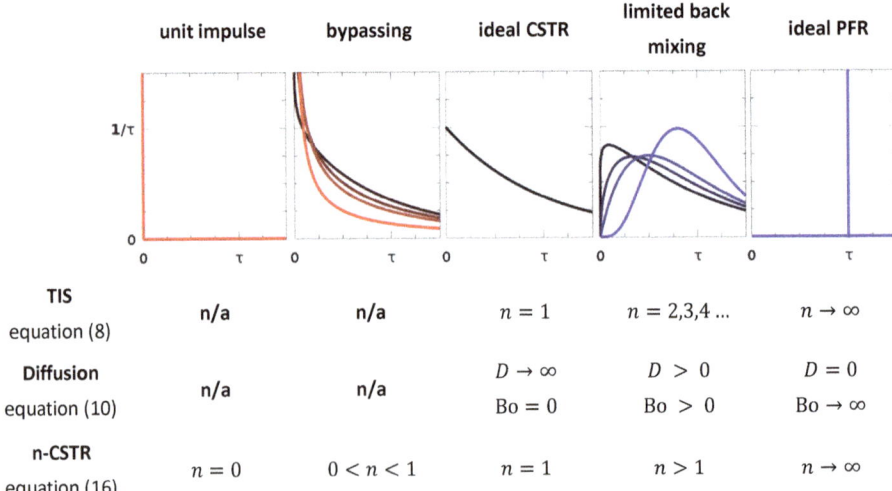

Figure 1. Comparison of the parameter range of tanks-in-series (TIS), diffusion, and generalised cascade of n continuous stirred tank reactor (n-CSTR) models.

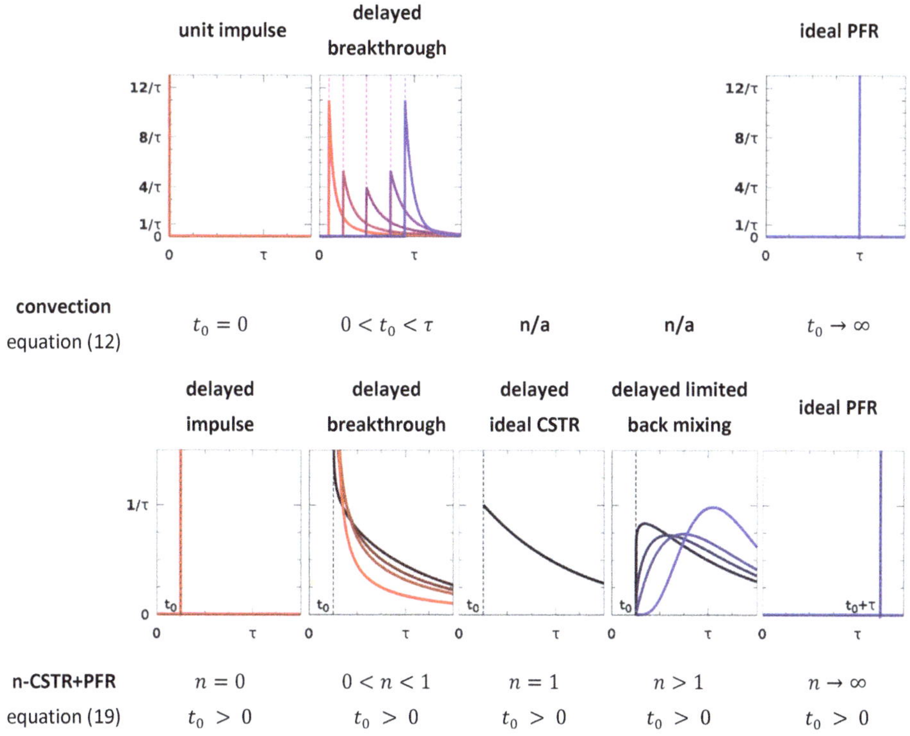

Figure 2. Comparison of the parameter range of n-CSTR+PFR (combined n-CSTR + plug flow reactor) model with the convection model.

2. Fundamentals of RTD Modelling

The first step in RTD modelling is obtaining the residence time distribution. The experimental setup uses a small amount of tracer material added to the process. The concentration of the tracer material is measured at the relevant outlet. The result is the tracer concentration profile $C(t)$. Once the tracer concentration has decayed completely, the concentration profile is normalised to the so-called $E(t)$ of the RTD:

$$E(t) = \frac{C(t)}{\int_0^\infty C(t) \cdot \mathrm{d}t} \tag{1}$$

The $E(t)$ curve describes the distribution of exit times—its peaks indicate the time where most of the tracer material is discharged. As $E(t)$ is a normalised distribution with integral 1, it is possible to calculate statistical indicators (mean residence time τ, standard deviation σ) directly from the $E(t)$ curve:

$$\begin{aligned} \tau &= \int_0^\infty t \cdot E(t) \cdot \mathrm{d}t \\ \sigma^2 &= \int_0^\infty (t-\tau)^2 \cdot E(t) \cdot \mathrm{d}t \end{aligned} \tag{2}$$

The key to RTD modelling is the ability to combine multiple RTDs—multiple $E(t)$ curves—into one process-level RTD. If a process with an RTD $E_1(t)$ feeds into a process with a different RTD $E_2(t)$, the combined RTD is calculated by the convolution integral

$$\begin{aligned} \mathrm{RTD}(t) &= \int_{\theta=0}^{\theta=t} E_1(t-\theta) \cdot E_2(\theta) \cdot \mathrm{d}\theta \\ &= \int_{\theta=0}^{\theta=t} E_1(\theta) \cdot E_2(t-\theta) \cdot \mathrm{d}\theta \end{aligned} \tag{3}$$

A shorthand notation for convolution uses the * operator. The convolution is symmetric:

$$\mathrm{RTD}(t) = (E_1 * E_2)(t) = (E_2 * E_1)(t) \tag{4}$$

The convolution operation has a neutral element—the Dirac (or unit) impulse $\delta(t)$. If any distribution $E(t)$ is convolved with the Dirac impulse, the result is the distribution itself, as shown in Equation (5). Tracer experiments are the practical application of this equation: if a tracer impulse is added at the inlet, the RTD of the process can be observed as a tracer concentration profile at the outlet. Figure 1 shows the unit impulse in red.

$$(E * \delta)(t) = E(t) \tag{5}$$

3. RTD Models and Their Limits

3.1. Ideal Plug Flow Reactor (PFR)

The ideal continuous stirred tank reactor (CSTR) and the ideal plug flow reactor (PFR) are the most basic models for continuous reactors and are widely used model prototypes in the chemical engineering community [16,17]. Both are characterised by their mean residence time τ. If a tracer pulse is added at the inlet of the ideal PFR, the response at the outlet is a delayed pulse of the same shape. Therefore, the residence time distribution of the ideal PFR is a Dirac impulse delayed by τ—see Equation (6). The RTD of the ideal PFR is shown in Figure 1 in blue.

$$\mathrm{RTD}_{\mathrm{PFR},\tau}(t) = \delta(t-\tau) \tag{6}$$

3.2. Ideal Continuous Stirred Tank Reactor (CSTR)

The ideal CSTR is characterised by perfect back mixing. Each portion of the material has the same chance to be discharged at the outlet, regardless of how long it has already been inside the CSTR. If some tracer is added to the CSTR, it is perfectly mixed within the tank and a portion of the tracer material is immediately visible at the outlet. The constant discharge chance leads to an exponentially decaying tracer profile. The assumption of perfect mixing directly leads to the typical exponential RTD shape of the ideal CSTR (black line in Figure 1). The RTD of an ideal CSTR with mean residence time τ is given by:

$$\mathrm{RTD}_{\mathrm{CSTR},\tau}(t) = \frac{1}{\tau}\exp\left\{-\frac{t}{\tau}\right\} \tag{7}$$

3.3. Tanks-in-Series (TIS)

A popular model for non-ideal CSTRs is the cascaded CSTR or tanks-in-series (TIS) model. A defined number of CSTRs (n) are chained together such that the inlet of one CSTR is connected to the inlet of the next. Due to the fact that material can only flow from one CSTR in the cascade to the next, but not backwards, the TIS model describes imperfect back mixing. Each CSTR in the chain has the same mean residence time of τ/n. The RTD of the TIS model can be calculated by convolving the RTD of the CSTRs n times with itself. MacMullin and Weber [15] were the first to derive a general formula for the TIS model with n CSTRs in series:

$$\mathrm{RTD}_{n,\tau}(t) = \frac{t^{n-1}}{(n-1)!}\left(\frac{n}{\tau}\right)^n \exp\left\{-\frac{t\,n}{\tau}\right\} \tag{8}$$

For $n = 1$, Equation (8) reduces to Equation (7), the RTD of an ideal CSTR. With higher values of n, the peak of the RTD moves from $t = 0$ closer to $t = \tau$, and the height of the peak increases.

The theoretical limit for $n \rightarrow \infty$ is a singular peak at $t = \tau$, which is the RTD of an ideal PFR with the same mean residence time τ:

$$\mathrm{RTD}_{n\rightarrow\infty,\tau}(t) \rightarrow \delta(t-\tau) = \mathrm{RTD}_{\mathrm{PFR},\tau}(t) \quad (9)$$

Figure 1 shows the RTD shapes and transition from ideal CSTR, over imperfect back-mixing, to PFR as a black–blue gradient.

3.4. Diffusion Model

A classical model for non-ideal plug-flow reactors is the dispersion model. The basis for this is the dimensionless partial differential equation:

$$\frac{\partial C}{\partial \theta} = \left(\frac{D}{uL}\right) \cdot \frac{\partial^2 C}{\partial x^2} - \frac{\partial C}{\partial x} \quad (10)$$

with the concentration $C = C(x,\theta)$ as a function of non-dimensional position in the tube x and non-dimensional time $\theta = t/\tau$, the axial diffusion coefficient D, the velocity of the material along the tube u, and the length of the tube L.

The residence time distribution is then calculated from the concentration over time at the end of the tube ($x = 1$). The exact analytical solution for closed boundary conditions is unknown. There are approximations for low diffusion $(D/uL) < 0.01$, where the RTD is a Gaussian distribution centred at the mean residence time τ and variance $\sigma^2 = (D/uL)$. For higher levels of diffusion, the results have to be obtained numerically [16,18,20]. The results show that as the diffusion in the non-ideal increases, the RTD assumes more and more the shape of the TIS model with low values of n. As $D \rightarrow \infty$, the solutions approach the exponential distribution of the ideal CSTR.

In a sense, the CSTR and PFR reactor models are mirror images: the worst possible CSTR imaginable with no back mixing at all ($n \rightarrow \infty$) is the ideal PFR ($D = 0$), and conversely, the worst possible PFR with infinitely high diffusion ($D \rightarrow \infty$) has the perfect mixing properties of a CSTR ($n = 1$). The limits and RTD shapes are summarised in Figure 1.

3.5. Convection Model

For very viscous fluids or very short tubular reactors, the dispersion model for non-ideal plug-flow reactors may not feasible. In this case, the RTD is a consequence of the characteristic parabolic velocity profile of a laminar flow, not driven by diffusion. Ananthakrishnan et al. [21] derived an RTD for a pure convective flow with mean residence time τ.

$$\mathrm{RTD}_{\mathrm{conv},\tau}(t) = \tfrac{\tau^3}{2 \cdot t^3} \quad \text{if} \quad t \geq \tfrac{\tau}{2} \quad (11)$$

If the convection model has a dead time $t_0 = \tau/2$ before the first material exists the tube, followed by a t^{-3} decay. This effect is independent from the velocity of the flow. There are some generalisations to this model (for example, References [30,31]), but Gutierrez et al. [24] gave a dimensionless generalisation which is parametrised with the normalised breakthrough time $\theta_0 = t_0/\tau$ with $0 \leq \theta_0 \leq 1$:

$$\mathrm{RTD}_{\mathrm{conv},\theta_0}(\theta) = \tfrac{1}{1-\theta_0} \cdot \tfrac{1}{\theta} \cdot \left(\tfrac{\theta_0}{\theta}\right)^{\frac{1}{1-\theta_0}} \quad \text{if} \quad \theta \geq \theta_0 \quad (12)$$

The $\theta_0 = 0.5$ Equation (12) simplifies to a non-dimensional version of Equation (11). Equation (12) can also be rewritten to a dimensional version with mean residence time τ and breakthrough time t_0 with $0 \leq t_0 \leq \tau$:

$$\mathrm{RTD}_{\mathrm{conv},\tau,t_0}(t) = \tfrac{\tau^2}{t \cdot (\tau - t_0)} \cdot \left(\tfrac{t_0}{t}\right)^{\frac{\tau}{\tau - t_0}} \quad \text{if} \quad t \geq t_0 \quad (13)$$

An interesting feature of this parameterisation is that the two limiting cases are the unit impulse $\delta(t)$ as the breakthrough time approaches zero ($t_0 \rightarrow 0$) and a delayed Dirac impulse $\delta(t - \tau)$—the RTD of an ideal PFR—as the breakthrough time approaches the mean residence time ($t_0 \rightarrow \tau$). The RTD shapes and limits of the convection model are summarised in Figure 2, the transition from unit impulse to PFR is shown as a red–blue colour gradient.

4. Generalised Cascade of *n* Continuous Stirred Tank Reactors: The n-CSTR Model

The diffusion, TIS, and convection models cover a wide variety of RTD shapes. However, because of their different analytical forms, the only way to describe transitions between a bypassing condition and a non-ideal mixing condition, is to build reactor networks containing convective elements and CSTR cascades or utilising parallel CSTRs in reactor networks [13,25,28,32]. The TIS model has been generalised to a non-integer number of CSTRs in series [25], but it is also possible to utilise less than one CSTR to describe short-circuiting and bypassing effects [29]. The result is the generalised n-CSTR model.

4.1. The $\Gamma(n)$ Function

The basis for the generalised n-CSTR Model is the TIS model in Equation (8) which is limited to natural numbers of n. This limitation comes from the factorial expression $(n-1)!$ in Equation (8). If the factorial could be replaced by a real-valued function, it would be possible to provide an analytical residence time distribution of n CSTRs for *any* value of n. Such a function exists—the gamma function $\Gamma(n)$.

The gamma function traces back to Euler, who defined the first analytic continuation of the factorial [33]. The gamma function is defined with an absolute converging infinite integral:

$$\Gamma(n) = \int_0^{\infty} x^{n-1} \cdot e^{-n} \cdot dx \tag{14}$$

and suffices the following conditions:

$$\begin{aligned} &\Gamma(0) = \Gamma(1) = 1 \\ &n \cdot \Gamma(n) = \Gamma(n+1) \; \forall \, n \in \mathbb{R} \\ &\Gamma(n) = (n-1)! \; \forall \, n \in \mathbb{N} \end{aligned} \tag{15}$$

The gamma function is available in many programming languages and tools, for example C [34], Python via the SciPy library [35], Matlab [36], and Microsoft Excel [37]. With the help of the gamma function, it is possible to re-write Equation (8) to be defined for any positive n:

$$\mathrm{RTD}_{n,\tau}(t) = \frac{t^{n-1}}{\Gamma(n)} \cdot \left(\frac{n}{\tau}\right)^n \cdot \exp\left\{-\frac{t\,n}{\tau}\right\} \tag{16}$$

4.2. Influence of Shape Parameter n

The RTD of a generalised cascade in Equation (16) has two parameters: the mean residence time τ and the number of CSTRs in the cascade n. The generalisation allows a finer tuning of the shape of the RTD, because non-integer values of n are allowed. While a cascade of 1.5 CSTRs may seem counter-intuitive at first, non-integer values can occur when analysing experimental RTD curves. The textbook example is calculating n from the mean residence time and variance of a measured RTD:

$$n = \frac{\tau^2}{\sigma^2} \tag{17}$$

With the limitation to integer n in the standard TIS model, it is usually recommended to round n to the nearest integer [17,38]. With the generalised n-CSTR model it is possible to use the non-integer n directly, to fine-tune the shape of the RTD (Figure 3).

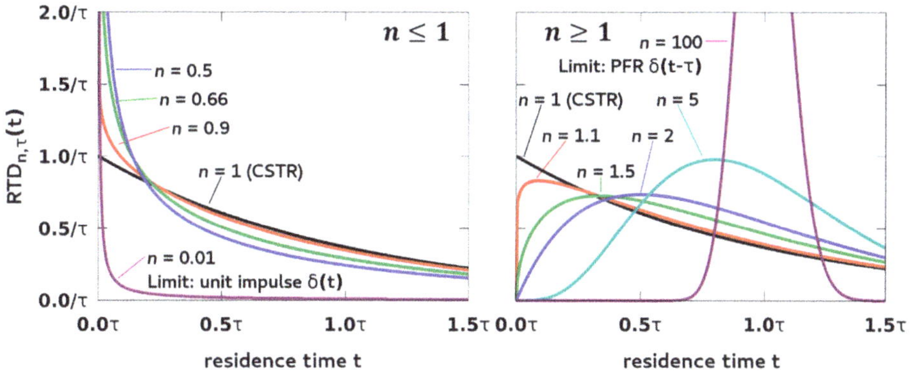

Figure 3. Influence of parameter n on the RTD shape for $n \leq 1$ and $n \geq 1$.

The number of n CSTRs is also connected to the Bodenstein number Bo, which is indirectly proportional to the diffusion in the system. Thus, calculating the Bodenstein number for a process and relating it back to the number of CSTRs in the cascade can result into a non-integer n. The relation between n and the Bodenstein number Bo is given by Elgeti [39]:

$$n = \frac{\mathrm{Bo}}{2} + 1 \tag{18}$$

As a side note, the Bodenstein number Bo and the Peclet number Pe are sometimes used interchangeably in this context, although the numbers have a slightly different meaning [16]. Just like the TIS model, the generalised n-CSTR cascade connects the ideal CSTR model ($n = 1$, Bo = 0, $D \rightarrow \infty$) and the ideal PFR model ($n \rightarrow \infty$, Bo$\rightarrow \infty$, $D = 0$) with a region of limited back mixing ($1 < n < \infty$).

The novelty of the n-CSTR model is expansion to the case $n < 1$. The resulting RTDs show a sharp peak at $t = 0$, while still maintaining the exponentially shaped tail. The initial peak describes a short-circuiting or bypassing behaviour without changing the overall mean residence time τ. This behaviour is impossible to describe with the classical TIS and diffusion models. A common practice to describe bypassing is building complex reactor networks utilising TIS or diffusion models with short residence times, e.g., [13,25].

A way to build intuition for a cascade with less than one CSTR is to consider a single CSTR, but it is not fully utilised and thus some material is able to bypass the mixing, moving directly from inlet to outlet. It makes sense to describe this scenario with $n < 1$. A real-world example is a vertical continuous mixing device described in Reference [29]. The construction is based on an ideal CSTR and for a range of operating conditions it behaves exactly like one ($n = 1$). Small deviations from these operating conditions cause small deviations from the ideal CSTR behaviour which fall either on the limited back-mixing side ($n > 1$) or on the increased initial peak side ($n < 1$). As it is possible to describe both non-ideal cases with the n-CSTR model with only one parameter, fitting the shape parameter n to the obtained RTDs provides a way to describe the quality of the process with a single number.

As n becomes smaller and smaller, the RTD peak becomes sharper and converges to the unit impulse $\delta(t)$ for $n \rightarrow 0$ (Figure 3). Intuitively, if there are no CSTRs in the cascade at all, there is nothing that could change the RTD of any incoming material. A visual comparison of the generalised n-CSTR model with the TIS and diffusion model is given in Figure 1.

The behaviour for $n < 1$ is similar to the convection model; however, the convection model adds a delay time t_0, whereas the peak stays at $t = 0$ in the generalised CSTR cascade. If needed, the n-CSTR model can be combined with an ideal PFR to add a delay time:

$$\begin{aligned} \mathrm{RTD}_{n,\tau,t_0}(t) &= \left(\mathrm{RTD}_{\mathrm{PFR},t_0} * \mathrm{RTD}_{n,\tau}\right)(t) \\ &= \mathrm{RTD}_{n,\tau}(t - t_0) \quad \text{if} \quad t \geq t_0 \end{aligned} \tag{19}$$

The mean residence time of this system is $\tau + t_0$. The parameter range and possible RTD shapes of this model is summarised and compared to the convection model in Figure 2.

4.3. *Quantification of Bypassing Material Fraction*

Any n-CSTR values in the range $0 < n < 1$ are considered to model bypassing of material. The fraction of material which bypasses the reactor/mixer increases for smaller values of n. In order to quantify the bypassing material fraction, the cumulative residence time distribution has been calculated by numerically integrating over Equation (16), up until a certain bypassing time threshold $t_{\max}$.

Figure 4 shows the bypassing material fractions for $t_{\max} = 0.01\ \tau$, $0.05\ \tau$, $0.1\ \tau$. By defining the bypassing fraction as the integral of the first part of the RTD, the ideal CSTR at $n = 1$ also shows a small bypassing fraction. This fraction corresponds to the small amount of the newly added material which is perfectly mixed and immediately visible at the outlet. The bypassing fraction in the initial peak of the RTD increases slowly for n values close to 1 and shows a steep increase for n values close to 0. The limiting case for $n \to 0$ is the Dirac impulse, where all material instantly leaves at $t = 0$.

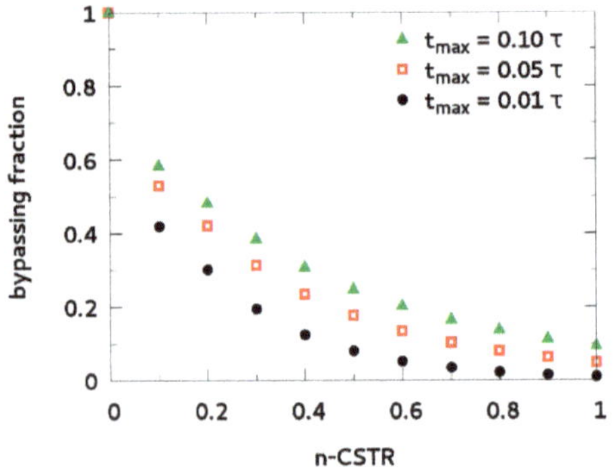

Figure 4. Bypassing mass fraction in the first 1%, 5%, and 10% of the mean residence time τ as function of the number of CSTRs n.

4.4. *Filtering of Mass Flow Fluctuations in a Continuous Manufacturing Line*

The RTD of the generalised n-CSTR model can in principle be used anywhere the models of the non-ideal CSTR or non-ideal PFR have been applied: description of complete processes, as a model for a single unit operation, or as a building block in a reactor network.

The idea of RTD modelling is to characterise each unit operation in the continuous manufacturing process (e.g., mixing, granulation, tableting) with the residence time distribution. The goal is to predict how the fluctuations in the initial material stream propagate throughout the process and effect the final product quality. If the RTD—in our case the generalised n-CSTR model $\mathrm{RTD}_{n,\tau}(t)$—and the input mass flow $\dot{m}_{in}(t)$ is known, it is possible to calculate the mass flow at the outlet $\mathrm{m}_{\mathrm{out}}(t)$ with a convolution integral [8]:

$$\dot{m}_{out}(t) = \left(\dot{m}_{in} * \mathrm{RTD}_{n,\tau}\right)(t) \tag{20}$$

Vertical continuous mixing devices have broad RTDs that filter out mass fluctuations in the input stream. Typical mean residence times in a continuous mixing device are in the range of 100 s, and good mixing behaviour is indicated with number of CSTRs n close to 1 [29].

Even if no data for $m_{in}(t)$ are available, it is possible to characterise the filterability of the RTD with Fourier analysis (for details see Reference [40]). The filterability Fe(f) indicates how much the frequency f is damped by the RTD. The filterability spectrum has been calculated using Matlab's built-in function fft() for calculating the fast Fourier transformation. The input is the RTD curve given by the n-CSTR model, with τ = 100 s in Equation (16). The RTD curve has been sampled with a constant time step Δt = 0.01 s up to 6000 s, thus the frequency spectrum is available up to 50 Hz with a resolution of 1/3000 Hz.

The filterability spectra are shown in Figure 5. All spectra start at one for 0 Hz, because it is impossible to filter a constant input stream. As the frequency increases, the amplitude of the fluctuations is reduced. For example, an RTD with n = 1 and τ = 100 s damps frequencies f = 1 Hz to Fe(f) $\approx$ 1/1000 of its initial value (see Figure 5a).

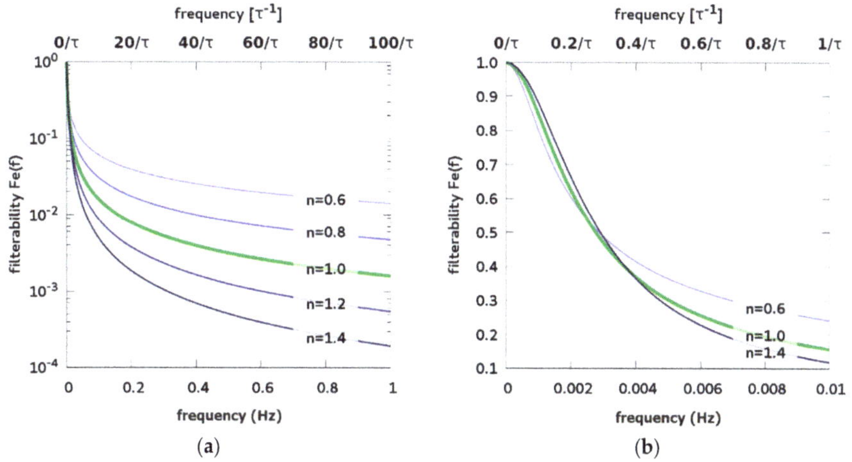

Figure 5. Filterability of a continuous mixing device with τ = 100 s and n-CSTR close to 1. (**a**) Higher values of n indicate better filtering of higher frequencies. (**b**) Very low frequencies below 0.3 τ^{-1} (i.e., drifts in the data that are longer than one third of the mean residence time τ) are better filtered with lower values of n-CSTR.

Generally speaking, residence time distributions with higher mean residence times are better frequency filters. The filterability spectrum is inversely proportional to the mean residence time τ. Increasing τ effectively compresses the filterability spectrum, reaching the low frequency amplitudes faster (see normalised frequencies τ^{-1} in Figure 5).

The value of n changes the shape of the frequency filter: Lower values of n filter very low frequencies f < $0.3\tau^{-1}$ slightly better (Figure 5b), but they also show significantly worse damping of higher frequencies (see Figure 5a). This can also be explained with the bypassing behaviour described by the n-CSTR model with n < 1: Rapid changes (high frequencies) in the inlet material stream are immediately visible at the outlet due to bypassing. However, slow drifts in the inlet stream (low frequency) are damped by the narrow and long tail of the remaining RTD after the initial peak.

Another way to analyse the filterability of the n-CSTR model is to construct an inlet condition and perform the convolution integral in Equations (3) and (20). Figure 6a shows the response of the n-CSTR model to a rectangular inlet condition with a length of one MRT. While appearing artificial, this rectangular inlet condition is a good model for set point changes occurring in a continuous manufacturing process [9,10,29]. Even though the change in the inlet condition (e.g., mass flow, concentration) persists for one mean residence time, the peak at the outlet is damped significantly compared to the height of the rectangle.

Figure 6b plots the outlet peak for different rectangle widths Δt and n-CSTR values close to 1. Higher n-CSTR values lead to better damping of the rectangular input and lower peaks at the outlet. This effect is more visible at narrower rectangle widths: the longer the rectangular inlet condition, the harder it is to dampen.

Figure 5b shows that lower n-CSTR values lead to a better damping behaviour at lower frequencies and higher durations, which seem to contradict the results from Figure 6b. The reason is that even rectangles with a high duration (and thus low frequency) have a complex frequency spectrum with high frequencies occurring at the flanks of the rectangle. Although lower n-CSTR values damp very low frequencies better, the filterability of high frequencies is significantly worse. The result is an overall higher response at low n-CSTR values, as shown in Figure 6b.

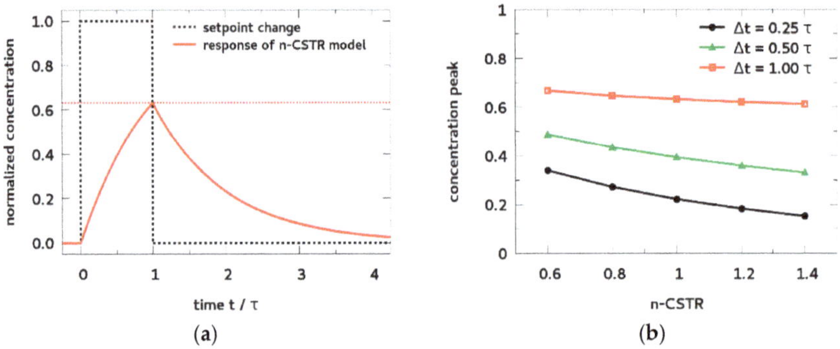

Figure 6. (**a**) Response of the n-CSTR model to a rectangle with a length $\Delta t = \tau$ and $n = 1$. (**b**) Concentration peak at the outlet for n-CSTR values around 1 for 3 different rectangle lengths $\Delta t = 0.25\,\tau, 0.5\,\tau, 1.0\,\tau$.

5. Conclusions

In this work, the tanks-in-series (TIS) model has been generalised to a cascade of an arbitrary number of continuous stirred tank reactors (n-CSTR model). Several unique properties of the n-CSTR model have been discussed.

The n-CSTR model does not only allow fine-tuning of the well-known TIS model with $n > 1$, but it also expands to a new class of residence time distributions describing short-circuiting and bypassing effects with $n < 1$. The already established convection model offers similarly shaped residence time distributions with a high initial peak for bypassing effects; however, the parameterisation with t_0 changes both shape and position of the peak simultaneously. Changing the number of CSTRs in the n-CSTR model changes only the shape of the distribution, but the start of the RTD always remains at $t = 0$. Thus, if desired, it is trivial to choose the starting position of the RTD of the n-CSTR model by applying an additional offset (t_0).

The n-CSTR model is the only model which connects the unit impulse ($n \to 0$), a bypassing regime ($0 < n < 1$), the ideal CSTR ($n = 1$), a limited back-mixing regime ($n > 1$), and the ideal plug flow reactor ($n \to \infty$) with the same analytical form for any given mean residence time τ, by only adjusting one shape parameter: n. Bypassing effects in processes with near ideal CSTR behaviour can be modelled with shape parameters $n < 1$, without the need for reactor networks with multiple fitting parameters. The smooth transition between bypassing and limited back-mixing enables a simple curve fitting with the shape parameter n. This parameter seconds as a descriptor for the mixing quality of a process, with values close to 1 being optimal (ideal CSTR).

Lastly, the applicability of the generalised n-CSTR model has been demonstrated by analysing bypassing fractions, dampening behaviour of fluctuations occurring in a continuous manufacturing line, and the response to set point changes.

Author Contributions: Conceptualisation: P.T., P.D., and D.J.; Methodology: P.T. and D.J.; Project administration: P.D.; Supervision: D.J.; Visualisation: P.T.; Writing—original draft: P.T.; Writing—review and editing: P.D. and D.J.

Funding: This research received no external funding.

Acknowledgments: The Research Center Pharmaceutical Engineering was funded by the Austrian COMET Program under the auspices of the Austrian Federal Ministry for Transport, Innovation and Technology (BMVIT), the Austrian Federal Ministry of Digital and Economic Affairs (BMDW), and by the Federal State of Styria (Styrian Funding Agency SFG). COMET is managed by the Austrian Research Promotion Agency FFG.

Conflicts of Interest: The authors declare no conflict of interest.

References

1. Plumb, K. Continuous Processing in the Pharmaceutical Industry. *Chem. Eng. Res. Des.* **2005**, *83*, 730–738. [CrossRef]
2. Schaber, S.D.; Gerogiorgis, D.I.; Ramachandran, R.; Evans, J.M.B.; Barton, P.I.; Trout, B.L. Economic Analysis of Integrated Continuous and Batch Pharmaceutical Manufacturing: A Case Study. *Ind. Eng. Chem. Res.* **2011**, *50*, 10083–10092. [CrossRef]
3. Allison, G.; Cain, Y.T.; Cooney, C.; Garcia, T.; Bizjak, T.G.; Holte, O.; Jagota, N.; Komas, B.; Korakianiti, E.; Kourti, D.; et al. Regulatory and Quality Considerations for Continuous Manufacturing May 20–21, 2014 Continuous Manufacturing Symposium. *J. Pharm. Sci.* **2015**, *104*, 803–812. [CrossRef] [PubMed]
4. Aigner, I.; Hsiao, W.-K.; Dujmovic, D.; Stegemann, S.; Khinast, J. Methodology for Economic and Technical Comparison of Continuous and Batch Processes to Enhance Early Stage Decision-making. In *Continuous Manufacturing of Pharmaceuticals*; Kleinebudde, P., Ed.; John Wiley & Sons, Ltd.: Chichester, UK, 2017; pp. 485–505. ISBN 978-1-119-00134-8.
5. Adam, S.; Suzzi, D.; Radeke, C.; Khinast, J.G. An integrated Quality by Design (QbD) approach towards design space definition of a blending unit operation by Discrete Element Method (DEM) simulation. *Eur. J. Pharm. Sci.* **2011**, *42*, 106–115. [CrossRef] [PubMed]
6. Yu, L.X. Pharmaceutical Quality by Design: Product and Process Development, Understanding, and Control. *Pharm. Res.* **2008**, *25*, 781–791. [CrossRef] [PubMed]
7. Almaya, A.; De Belder, L.; Meyer, R.; Nagapudi, K.; Lin, H.-R.H.; Leavesley, I.; Jayanth, J.; Bajwa, G.; DiNunzio, J.; Tantuccio, A.; et al. Control Strategies for Drug Product Continuous Direct Compression—State of Control, Product Collection Strategies, and Startup/Shutdown Operations for the Production of Clinical Trial Materials and Commercial Products. *J. Pharm. Sci.* **2017**, *106*, 930–943. [CrossRef] [PubMed]
8. Engisch, W.; Muzzio, F. Using Residence Time Distributions (RTDs) to Address the Traceability of Raw Materials in Continuous Pharmaceutical Manufacturing. *J. Pharm. Innov.* **2016**, *11*, 64–81. [CrossRef]
9. Rehrl, J.; Kruisz, J.; Sacher, S.; Khinast, J.; Horn, M. Optimized continuous pharmaceutical manufacturing via model-predictive control. *Int. J. Pharm.* **2016**, *510*, 100–115. [CrossRef]
10. Kruisz, J.; Rehrl, J.; Sacher, S.; Aigner, I.; Horn, M.; Khinast, J.G. RTD modeling of a continuous dry granulation process for process control and materials diversion. *Int. J. Pharm.* **2017**, *528*, 334–344. [CrossRef]
11. Martinetz, M.C.; Karttunen, A.-P.; Sacher, S.; Wahl, P.; Ketolainen, J.; Khinast, J.G.; Korhonen, O. RTD-based material tracking in a fully-continuous dry granulation tableting line. *Int. J. Pharm.* **2018**, *547*, 469–479. [CrossRef]
12. Liotta, F.; Chatellier, P.; Esposito, G.; Fabbricino, M.; Van Hullebusch, E.D.; Lens, P.N.L. Hydrodynamic Mathematical Modelling of Aerobic Plug Flow and Nonideal Flow Reactors: A Critical and Historical Review. *Crit. Rev. Environ. Sci. Technol.* **2014**, *44*, 2642–2673. [CrossRef]
13. Gorzalski, A.S.; Harrington, G.W.; Coronell, O. Modeling Water Treatment Reactor Hydraulics Using Reactor Networks. *J.-Am. Water Works Assoc.* **2018**, *110*, 13–29. [CrossRef]
14. Sheoran, M.; Chandra, A.; Bhunia, H.; Bajpai, P.K.; Pant, H.J. Residence time distribution studies using radiotracers in chemical industry—A review. *Chem. Eng. Commun.* **2018**, *205*, 739–758. [CrossRef]

15. MacMullin, R.; Weber, M. The theory of short-circuiting in continuous-flow mixing vessels in series and the kinetics of chemical reactions in such systems. *Trans. Am. Inst. Chem. Eng.* **1935**, *31*, 409–458.
16. Levenspiel, O. *Chemical Reaction Engineering*, 3rd ed.; Wiley: New York, NY, USA, 1999; ISBN 978-0-471-25424-9.
17. Fogler, H.S. *Elements of Chemical Reaction Engineering*, 4th ed.; Prentice Hall PTR International Series in the Physical and Chemical Engineering Sciences; Prentice Hall PTR: Upper Saddle River, NJ, USA, 2006; ISBN 978-0-13-047394-3.
18. Levenspiel, O.; Smith, W.K. Notes on the diffusion-type model for the longitudinal mixing of fluids in flow. *Chem. Eng. Sci.* **1957**, *6*, 227–235. [CrossRef]
19. Bischoff, K.B.; Levenspiel, O. Fluid dispersion-generalization and comparison of mathematical models—I generalization of models. *Chem. Eng. Sci.* **1962**, *17*, 245–255. [CrossRef]
20. Van der Laan, E.T. Notes on the diffusion-type model for the longitudinal mixing in flow. *Chem. Eng. Sci.* **1958**, *7*, 187–191.
21. Ananthakrishnan, V.; Gill, W.N.; Barduhn, A.J. Laminar dispersion in capillaries: Part I. Mathematical analysis. *AIChE J.* **1965**, *11*, 1063–1072. [CrossRef]
22. Ruthven, D.M. The residence time distribution for ideal laminar flow in helical tube. *Chem. Eng. Sci.* **1971**, *26*, 1113–1121. [CrossRef]
23. Levien, K.L.; Levenspiel, O. Optimal product distribution from laminar flow reactors: Newtonian and other power-law fluids. *Chem. Eng. Sci.* **1999**, *54*, 2453–2458. [CrossRef]
24. Gutierrez, C.G.C.C.; Dias, E.F.T.S.; Gut, J.A.W. Residence time distribution in holding tubes using generalized convection model and numerical convolution for non-ideal tracer detection. *J. Food Eng.* **2010**, *98*, 248–256. [CrossRef]
25. Martin, A.D. Interpretation of residence time distribution data. *Chem. Eng. Sci.* **2000**, *55*, 5907–5917. [CrossRef]
26. Mohammed, F.M.; Roberts, E.P.L.; Hill, A.; Campen, A.K.; Brown, N.W. Continuous water treatment by adsorption and electrochemical regeneration. *Water Res.* **2011**, *45*, 3065–3074. [CrossRef] [PubMed]
27. Dittrich, E.; Klincsik, M. Analysis of conservative tracer measurement results using the Frechet distribution at planted horizontal subsurface flow constructed wetlands filled with coarse gravel and showing the effect of clogging processes. *Environ. Sci. Pollut. Res.* **2015**, *22*, 17104–17122. [CrossRef] [PubMed]
28. Braga, B.M.; Tavares, R.P. Description of a New Tundish Model for Treating RTD Data and Discussion of the Communication "New Insight into Combined Model and Revised Model for RTD Curves in a Multi-strand Tundish" by Lei. *Metall. Mater. Trans. B* **2018**, *49*, 2128–2132. [CrossRef]
29. Toson, P.; Siegmann, E.; Trogrlic, M.; Kureck, H.; Khinast, J.; Jajcevic, D.; Doshi, P.; Blackwood, D.; Bonnassieux, A.; Daugherity, P.D.; et al. Detailed modeling and process design of an advanced continuous powder mixer. *Int. J. Pharm.* **2018**, *552*, 288–300. [CrossRef] [PubMed]
30. Heibel, A.K.; Lebens, P.J.M.; Middelhoff, J.W.; Kapteijn, F.; Moulijn, J. Liquid residence time distribution in the film flow monolith reactor. *AIChE J.* **2005**, *51*, 122–133. [CrossRef]
31. García-Serna, J.; García-Verdugo, E.; Hyde, J.R.; Fraga-Dubreuil, J.; Yan, C.; Poliakoff, M.; Cocero, M.J. Modelling residence time distribution in chemical reactors: A novel generalised n-laminar model. *J. Supercrit. Fluids* **2007**, *41*, 82–91. [CrossRef]
32. Leray, S.; Engdahl, N.B.; Massoudieh, A.; Bresciani, E.; McCallum, J. Residence time distributions for hydrologic systems: Mechanistic foundations and steady-state analytical solutions. *J. Hydrol.* **2016**, *543*, 67–87. [CrossRef]
33. Davis, P.J. Leonhard Euler's Integral: A Historical Profile of the Gamma Function: In Memoriam: Milton Abramowitz. *Am. Math. Mon.* **1959**, *66*, 849. [CrossRef]
34. Free Software Foundation The GNU C Library: Special Functions. Available online: http://www.gnu.org/software/libc/manual/html_node/Special-Functions.html (accessed on 23 January 2018).
35. The Scipy Community Scipy.Special.Gamma—SciPy v0.14.0 Reference Guide. Available online: https://docs.scipy.org/doc/scipy-0.14.0/reference/generated/scipy.special.gamma.html (accessed on 23 January 2018).
36. Mathworks Inc. Gamma Function-MATLAB Gamma-MathWorks. Available online: https://mathworks.com/help/matlab/ref/gamma.html?requestedDomain=true (accessed on 23 January 2018).

37. Microsoft Corporation GAMMA Function. Available online: https://support.office.com/en-us/article/gamma-function-ce1702b1-cf55-471d-8307-f83be0fc5297 (accessed on 23 January 2018).
38. Mo, Y.; Jensen, K.F. A miniature CSTR cascade for continuous flow of reactions containing solids. *React. Chem. Eng.* **2016**, *1*, 501–507. [CrossRef]
39. Elgeti, K. A new equation for correlating a pipe flow reactor with a cascade of mixed reactors. *Chem. Eng. Sci.* **1996**, *51*, 5077–5080. [CrossRef]
40. Gao, Y.; Muzzio, F.; Ierapetritou, M. Characterization of feeder effects on continuous solid mixing using fourier series analysis. *AIChE J.* **2011**, *57*, 1144–1153. [CrossRef]

processes

MDPI

Article

Dynamic Flowsheet Model Development and Sensitivity Analysis of a Continuous Pharmaceutical Tablet Manufacturing Process Using the Wet Granulation Route

Nirupaplava Metta [1,†], Michael Ghijs [2,3,†], Elisabeth Schäfer [4], Ashish Kumar [4], Philippe Cappuyns [4], Ivo Van Assche [4], Ravendra Singh [1], Rohit Ramachandran [1], Thomas De Beer [3], Marianthi Ierapetritou[1,*] and Ingmar Nopens[2,*]

1 Department of Chemical and Biochemical Engineering, Rutgers, The State University of New Jersey, Piscataway, NJ 08854, USA; nirupa.metta@rutgers.edu (N.M.); rs1034@scarletmail.rutgers.edu (R.S.); rohitrr@soe.rutgers.edu (R.R.)

2 BIOMATH, Department of Data Analysis and Mathematical Modelling, Ghent University, B-9000 Ghent, Belgium; Michael.Ghijs@UGent.be

3 Laboratory of Pharmaceutical Process Analytical Technology, Ghent University, B-9000 Ghent, Belgium; Thomas.DeBeer@UGent.be

4 Discovery, Product Development & Supply, Janssen, Pharmaceutical Companies of Johnson & Johnson, 2340 Beerse, Belgium; eschaef5@ITS.JNJ.com (E.S.); akuma328@its.jnj.com (A.K.); PCAPPUYN@its.jnj.com (P.C.); IVASSCHE@its.jnj.com (I.V.A.)

* Correspondence: marianth@soe.rutgers.edu (M.I.); Ingmar.Nopens@UGent.be (I.N.)

† These authors contributed equally to this work.

Received: 12 March 2019 ; Accepted: 15 April 2019 ; Published: 24 April 2019

Abstract: In view of growing interest and investment in continuous manufacturing, the development and utilization of mathematical model(s) of the manufacturing line is of prime importance. These models are essential for understanding the complex interplay between process-wide critical process parameters (CPPs) and critical quality attributes (CQAs) beyond the individual process operations. In this work, a flowsheet model that is an approximate representation of the ConsiGma^{TM}-25 line for continuous tablet manufacturing, including wet granulation, is developed. The manufacturing line involves various unit operations, i.e., feeders, blenders, a twin-screw wet granulator, a fluidized bed dryer, a mill, and a tablet press. The unit operations are simulated using various modeling approaches such as data-driven models, semi-empirical models, population balance models, and mechanistic models. Intermediate feeders, blenders, and transfer lines between the units are also simulated. The continuous process is simulated using the flowsheet model thus developed and case studies are provided to demonstrate its application for dynamic simulation. Finally, the flowsheet model is used to systematically identify critical process parameters (CPPs) that affect process responses of interest using global sensitivity analysis methods. Liquid feed rate to the granulator, and air temperature and drying time in the dryer are identified as CPPs affecting the tablet properties.

Keywords: model integration; flowsheet modeling; sensitivity analysis; continuous manufacturing; wet granulation

1. Introduction

Flowsheet models are approximate mathematical representations of the manufacturing line. The incentives for flowsheet model development for the pharmaceutical industry have been described in the paper of Escotet-Espinoza et al. [1]. Gernaey et al. [2] also wrote extensively on the value of

Process Systems Engineering (PSE) for pharmaceutical process development, in which flowsheet models contribute to combine knowledge and models of different unit operations and different scales for a holistic understanding of the process. The incentives for flowsheet modeling boil down to the in-silico achievement of process design and optimization, control system design and optimization, and an accurate risk assessment tool that could be used for regulatory instances. The first step in attaining these benefits is by the development of a flowsheet model that captures the relevant mechanisms for assessing the desired product properties as a function of process settings and material properties. This foundation built in this work, comprised of several diverse unit operation models that capture critical mechanisms, as a function of the process inputs.

The flowsheet model simulation enables the assessment of unit operation outputs downstream in the process. This has the advantage that, instead of applying the unit operation models separately for model-based research, the flowsheet model allows for targeted optimization of unit operation performance as a part of the entire line. In terms of the Quality-by-Design (QbD) paradigm, a flowsheet model allows for investigating the influence of critical process parameters (CPPs) in one unit to the critical quality attributes (CQAs) of material downstream in the process line. Namely, through flowsheet model development, process phenomena are directly linked to the final product quality downstream. Moreover, analysis of the developed model allows assessment of criticality of the various critical process parameters (CPPs) of the process, and subsequently research effort can be targeted towards those most critical areas.

In this work, composing the integrated system requires extensive synchronization of the unit operation models themselves, as these need to run seamlessly in one simulation, regardless of the different time scales, variable magnitudes, or stiffness of the various models. Population balance models, for example, require specific solution methods, and these need to run at par with other less computationally demanding models. This work therefore captures the research into the simultaneous simulation of these diverse models. In addition, process dynamics are included into the flowsheet model. This allows tracing the material properties throughout the entire line. This feature is included as the foundation of applying the model to assess the propagation of process disturbances, with respect to the product quality, i.e., which products needs to be discarded, or how fast can the process recover and return to a position where product critical quality attributes (CQAs) are within specification limits.

Previous work on the development of flowsheet models are restricted to direct compaction [3–6] and dry granulation routes [7,8] for continuous solid oral dosage manufacturing. Park et al. [7] created a flowsheet model of continuous dry granulation and applied it for optimization. Boukouvala et al. [9] developed a flowsheet model for the wet granulation route. This model served as a proof-of-concept and with no connection to specific experimental data. Boukouvala and Ierapetritou [10] also investigated a methodology for optimization of computationally expensive flowsheet models. Rogers and Ierapetritou [11] showed a flowsheet modeling case with hybrid models incorporating information from both detailed and reduced-order models.

The work presented in this manuscript includes models systematically developed based on experiments on units in the ConsiGma^{TM}-25 line for continuous tablet manufacturing using the same formulation and relevant materials across all the units. Specifically, the units involved are feeders, blenders, twin-screw wet granulator (TSWG), fluid bed dryer (FBD), comill, and tablet press. Models for these processes are developed [12–16] and included. Besides these, models for intermediate feeding and blending operations are also included. Transfer lines that lead to material holdup in between the units are added to the flowsheet model as well.

1.1. Objectives

The specific objectives of this work are:

- Develop a flowsheet model approximating the ConsiGma^{TM}-25 wet granulation manufacturing line;
- Demonstrate the use of the flowsheet model for simulating effects of disturbances in the continuous process;

- Identify critical process parameters (CPPs) affecting the properties of intermediate and final product.

Section 2.1 details various models used to build the flowsheet model. To enunciate how the flowsheet model can be used for propagation of information and disturbances across the units, a detailed discussion is provided in Section 3.1 along with supporting case studies in Section 3.2. In addition to building the flowsheet model, a detailed analysis of the developed model is provided. The scenario analysis, explained in Sections 2.2 and 3.3, serves to ensure that the flowsheet model which is a complex set of equations from various modeling approaches, runs successfully at several values of process variables and the process responses thus obtained are aligned with process knowledge. Following this, critical process parameters (CPPs) that affect product quality are identified through implementation of sensitivity analysis as explained in Sections 2.3 and 3.4.

2. Materials and Methods

The model formulation consisted of two active pharmaceutical ingredients (API), a lubricant and four excipients. Hereafter, the two two active pharmaceutical ingredients (API)s and the four excipients are referred to as API 1, API 2, and Excipient A, B, C, and D respectively. The formulation was processed using demineralized water as granulation liquid. The formulation used in this work is given in Table 1.

Table 1. Formulation used for model development.

Component Name	Weight %
API 1	75.58
API 2	8.72
Lubricant	0.58
Excipient A	6.05
Excipient B	1.51
Excipient C	6.05
Excipient D	1.51

2.1. Flowsheet Modeling

Since flowsheet models are approximate representations of the integrated manufacturing line, developing individual unit operation models aid in the development of a flowsheet model. In this work, the models developed are based on experiments conducted using the ConsiGma™-25 system (GEA Pharma systems, Collette, Wommelgem, Belgium), which is an oral solid dosage manufacturing line based on continuous wet granulation. In Sections 2.1.1–2.1.8 the individual unit models used in this work are briefly described. Figure 1 pictorially shows transfer of information across the unit operations i.e., feeder, blender, granulator, dryer, mill, and tablet press, in that order. Further, intermediate units are added and relevant information is transferred. These models are implemented in the software gPROMS FormulatedProducts v1.2.1 (PSE, London, UK). It is a platform for flowsheet simulations that uses an equation-oriented approach. An overview on the equation-oriented techniques particular to the gPROMS platform is given in Pantelides et al. [17].

2.1.1. Feeder

Loss-in-weight (LIW) feeders are used to feed the required powder components in the continuous manufacturing line. The feeder used in this work has a hopper and a conveying unit, a refill unit, and a PID controller. The hopper is used as a receptacle for the raw materials whereas the conveying system has a rotating screw that is used to move the material out of the feeder. The refill unit is used to feed material into the hopper when the fractional fill level in the hopper drops below a setpoint value. The PID controller enables the feeder to run in gravimetric mode, i.e., the screw speed in the conveying unit is adjusted to maintain a constant mass flow rate out of the unit. The three models (refill, feeding, and PID controller) work in conjunction to represent the overall feeding operation in the continuous line.

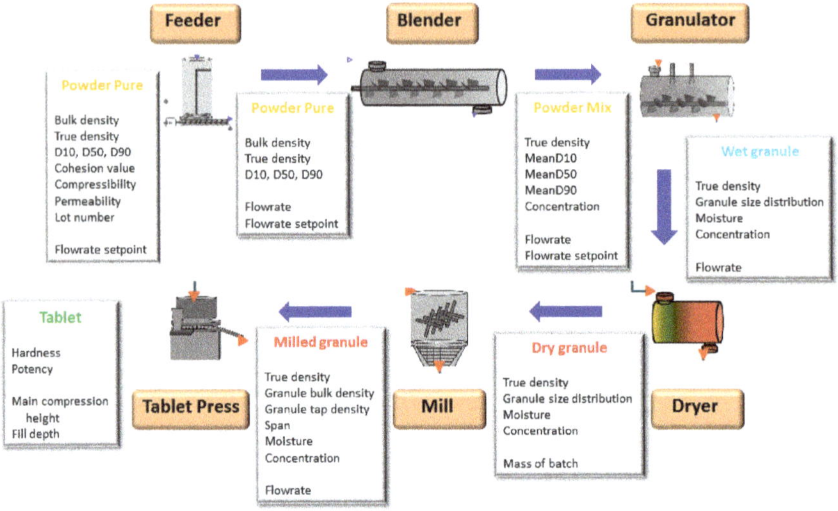

Figure 1. Schematic showing transfer of information between units required for flowsheet model development.

The mass flow rate out of the feeder is simulated using a feed factor model [3]. Feed factor is a time-dependent property ($ff(t)$), defined as maximum mass of powder fitting in a screw flight and has the unit of mass per screw revolution. It is found to be dependent on the amount of material in the hopper ($w(t)$) as given in Equation (1). The parameters ff_{max}, ff_{min}, and β are dependent on the powder bulk density, compressibility, cohesion and permeability. More details on the feed factor model and its dependence on material properties are published in [18]. The mass flow rate of the powder out of the feeder can then be obtained as given in Equation (2) where $\omega(t)$ is the screw speed that is manipulated by the PID controller.

$$ff(t) = ff_{max} + (ff_{min} - ff_{max})exp(-\beta w(t)) \tag{1}$$

$$\dot{M}_{out}(t) = ff(t)\omega(t) \tag{2}$$

2.1.2. Blender

Continuous blenders that are used to mix the powder components in the continuous line, dampen the flow rate variations from the feeding units. The build-up of mass in the blender $M(t)$ was found to be following a first order relationship as given in Equation (3), where M_{ss} is the steady state mass holdup and τ is the time constant. From the mass holdup and flow rate of the material into the blender ($\dot{M}_{in}$), flow rate out of the unit ($\dot{M}_{out}$) can be computed as given in Equation (4). An axial dispersion equation [3] is used to model the mixing calculation in the blending unit as a function of time. The equation as given in (5) is subject to initial and boundary conditions (Equation (6)). The coefficients of the axial dispersion model (τ_{ax} and Pe) are calculated based on their relationship to a CSTR-in-series model constant i.e., number of tanks n_t as given in Equation (7). Experimental data were used to develop regression models that predict the model constants τ, n_t, and M_{ss} as a function of flow rate and blade speed. More details on the blender model are available in [18].

$$\tau\frac{dM(t)}{dt} + M(t) = M_{SS} \tag{3}$$

$$\frac{dM(t)}{dt} = \dot{M}_{in} - \dot{M}_{out} \tag{4}$$

$$\tau_{ax}\frac{dC^i_{out}}{dt} = \frac{1}{Pe}\frac{\partial^2 C^i_{out}}{\partial \xi^2} - \frac{\partial C^i_{out}}{\partial \xi} \tag{5}$$

$$\begin{aligned} I.C: C^i_{out} &= 0, t = 0 \\ B.C: C^i_{out} &= C^i_{in}, \xi = 0 \\ \frac{dC^i_{out}}{dt} &= 0, \xi = 1 \end{aligned} \tag{6}$$

$$Pe = n_t + (8n_t + n_t^2)^{1/2} \tag{7}$$

2.1.3. Twin-Screw Wet Granulator

The modeling of the change in particle size distribution (PSD) of the material through granulation is executed with the compartmental population balance model of Van Hauwermeiren et al. [13]. The twin-screw wet granulator (TSWG) model mathematically links the aggregation and breakage behavior in the granulator barrel to the granulator process settings of mass flow rate, screw speed, and liquid flow rate. It moreover distinguishes two compartments in the barrel: the wetting zone (i.e., the zone where the liquid is added to the dry powder blend) where only aggregation occurs, followed by the kneading zone (i.e., kneading elements are present in the screws) with different aggregation behavior complimented with breakage. Each compartment is thus modeled by its own population balance model (PBM).

The PBM equation is given in Equation (8). The change in number of particles n of a certain size x over time t is thereby described based on aggregation kernel $\beta(t, x, \varepsilon)$, breakage selection function $S(\varepsilon)$ and breakage fragment distribution $b(x, \varepsilon)$. The aggregation kernel $\beta(t, x, \varepsilon)$ can be modeled as the product of aggregation efficiency β_0 with collision frequency $\beta(x, \varepsilon)$, as the relation is in this case independent of time t.

The formula of the collision frequency in the first PBM, describing the wetting zone, is given in Equation (9). It comprises a two-dimensional stepping function and a product kernel in order to reach bimodal granule (PSDs) starting from a monomodal powder (PSD). Kernel parameters β_0 , R_1, R_2, top $_1$, top $_2$, δ_1, and δ_2 are needed to achieve this mathematical connection [13].

$$\begin{aligned} \frac{\delta n(t,x)}{\delta t} =& \frac{1}{2}\int_0^x \beta(t, x-\varepsilon, \varepsilon) n(t, x-\varepsilon) n(t,\varepsilon) d\varepsilon \\ &- n(t,x)\int_0^\infty \beta(t,x,\varepsilon) n(t,\varepsilon) d\varepsilon \\ &+ \int_x^\infty b(t,x,\varepsilon) S(t,\varepsilon) n(t,\varepsilon) d\varepsilon - S(t,x) n(t,x) \end{aligned} \tag{8}$$

$$\begin{aligned} \beta(x,\varepsilon) = & \left(\frac{\text{top}_1}{2}\left(1 + \tanh\left(\frac{R_1^3 - \left(x^2+\varepsilon^2\right)^{1/2}}{\delta_1}\right)\right) \right. \\ & \left. - \frac{\text{top}_1 - \text{top}_2}{2}\left(1 + \tanh\left(\frac{R_2^3 - \left(x^2+\varepsilon^2\right)^{1/2}}{\delta_2}\right)\right)\right) \\ & \cdot \left(x^{1/3} \cdot \varepsilon^{1/3}\right) \end{aligned} \tag{9}$$

The breakage in the kneading zone is modeled by a linear breakage selection function:

$$S(\varepsilon) = S_0 \varepsilon^{1/3}, \tag{10}$$

and a breakage fragment distribution $b(x, \varepsilon)$ describing a combination of erosion and uniform breakage:

$$b(x,\varepsilon) = f_{prim}\frac{1}{\sqrt{2\pi}\sigma}e^{-\frac{\left(x^{1/3}-\mu\right)^2}{2\sigma^2}}\frac{\varepsilon}{\mu^3}\frac{1}{3x^{\frac{2}{3}}} + \left(1-f_{prim}\right)\frac{2}{\varepsilon}, \tag{11}$$

with ε the volume of the breaking particle, x the volume of the fragment, S_0 the breakage rate constant, σ and μ respectively the standard deviation and mean of the Gaussian distribution representing the size distribution of the small eroded particles, and f_{prim} the volume fraction of erosion in the overall breakage (as opposed to a fraction $(1 - f_{prim})$ of uniform breakage).

Aggregation in this zone could be described by a sum kernel (Equation (12)).

$$\beta(x,\varepsilon) = x + \varepsilon \tag{12}$$

Overall, kernel parameters R_2, β_0, and top_1 in the wetting zone and β_0, S_0, f_{prim}, μ, and σ in the kneading zone are linearly related to the process setting values in the units given in Table 2.

Table 2. Process setting ranges of the validated twin-screw wet granulator (TSWG) model [13].

Process Setting	Lower Bound	Upper Bound
Mass flow rate (kg/h)	10	20
Screw speed (RPM)	500	900
Liquid/solid-ratio (kg/kg)	0.08	0.18

2.1.4. Dryer

The fluidized bed dryer (fluid bed dryer (FBD)) model consists of prediction of granule batch drying kinetics based on single granule drying kinetics for one dryer cell [15]. The single granule drying kinetics are governed by Stefan diffusion of water vapor through the granule pores, from the source of evaporation to the edge of the granule. The mass transfer rates are corrected with the equilibrium moisture content X_e. When the moisture content of the granule is larger than its pore fraction, the remaining liquid is modeled as a layer of water of uniform thickness around the granule, evaporating according to a droplet:

$$\dot{m}_v = h_D(\rho_{v,s} - \rho_{v,\infty})A_d \tag{13}$$

with mass transfer rate $\dot{m}_v$ (kg s^{-1}), mass transfer coefficient h_D (m s^{-1}), partial vapor density over the droplet surface $\rho_{v,s}$ (kg m^{-3}), partial vapor density in the ambient air $\rho_{v,\infty}$ (kg m^{-3}), and droplet surface area A_d (m^2). The energy balance paired with this drying behavior is described by:

$$h_{fg}\dot{m}_v + c_{p,w}m_d\frac{dT_d}{dt} = h(T_g - T_d)4\pi R_d^2 \tag{14}$$

with specific heat of evaporation h_{fg} (J kg^{-1}), specific heat capacity of the liquid $c_{p,w}$ (J kg^{-1} K^{-1}), droplet mass m_d (kg), uniform droplet temperature T_d (K), heat transfer coefficient h (W m^{-2} K^{-1}), drying gas temperature T_g (K), and droplet radius R_d (m). After this layer of water is depleted the wet granule enters the subsequent drying phase. Herein the moisture is conceptualized as a sphere with radius R_i (m), also referred to as the wet core, filling up the pore volume of the granule with radius R_p (m). The mass transfer rate $\dot{m}_v$ in this stage is given by [19]:

$$\dot{m}_v = -\frac{8\pi\epsilon^{\beta}D_{v,cr}M_w p_g}{\Re(T_{cr,s} + T_{wc,s})}\frac{R_pR_i}{R_p - R_i}ln[\frac{p_g - p_{v,i}}{p_g - (\frac{\Re}{4\pi M_w h_D R_p^2}\dot{m}_v + \frac{p_{v,\infty}}{T_g})T_{p,s}}] \tag{15}$$

with ϵ the granule porosity (-), β an empirical coefficient, $D_{v,cr}$ the vapor diffusion coefficient (m^2 s^{-1}), M_w the liquid molecular weight (kg mol^{-1}), p_g the pressure of the drying air (Pa), $\Re$ the ideal gas constant (J mol^{-1} K^{-1}), $T_{cr,s}$ and $T_{wc,s}$ respectively the temperature of solids at the granule surface and at the gas-liquid interface (K), $p_{v,\infty}$ and $p_{v,i}$ respectively the partial vapor pressure in the drying air (Pa) and at the gas-liquid interface, and $T_{p,s}$ the temperature of the particle solids (K). Equation (14) is assumed to apply for the energy balance of the granule during this drying phase. The physical properties of the solids were assumed to be at environment conditions of 25 °C and atmospheric air pressure in the model, whereas liquid properties were modeled as those of pure water. A value of 35 % was assumed for the granule porosity. Finally, the mass transfer rates $\dot{m}_v$ are corrected with the effect of X_e:

$$\dot{m}_{v,res} = \frac{X - X_e}{X}\dot{m}_v \tag{16}$$

Overall this means that the course of moisture content X of a granule amounts to:

$$\dot{X}_{SPDM} = \frac{\dot{m}_{v,res}}{m_p} \tag{17}$$

with $\dot{X}_{SPDM}$ the change in moisture content X over time and m_p the total mass of the granule (particle). This is solved from time $t = 0$ until the FBD cell drying time $t = t_{dry}$.

Connecting the single granule drying kinetics to those of the batch, along with the continuous filling of the batch, is done according to the following simplified approach. Drying curves are calculated for several size fractions of the granules, in which the arithmetic mean size is representative in the X_{SPDM}. Thus the average moisture content $\overline{X}_f$ per size fraction f then equals the average moisture content of different drying curves, starting a certain time τ later over the cell filling time t_{fill}:

$$\overline{X}_f = \frac{\sum_{\tau=0}^{t_{fill}} X_{SPDM}(t-\tau)}{n_\tau}, \tag{18}$$

with n_τ the amount times τ that the drying curve was shifted over the cell filling time interval t_{fill}.

A constant ideal fluidization behavior, constant relative air humidity in the dryer cell and atmospheric air properties in the drying chamber (with exception of the drying agent temperature) were assumed in the batch approach.

The dryer model discussion so far dealt with the drying behavior of the material, other material properties are directly governed by the dynamic output of the TSWG model. For each dryer cell, these are mass-averaged over the drying cell filling period. This is illustrated in Equation (19) for the concentration of active pharmaceutical ingredients (API) in the dryer cell $C_{API,dryer}$, and is calculated the same way for the material true density, PSD and the mass of the material in the dryer cell. They are weighted by the mass flow rate at which they are flowing from the TSWG at time t ($MFR_{TSWG}(t)$).

$$C_{API,dryer} = \frac{\int_t^{t+t_{fill}} C_{API,TSWG}(u)MFR_{TSWG}(u)du}{\int_t^{t+t_{fill}} MFR_{TSWG}(u)du} \tag{19}$$

After drying, the breakage of the material through pneumatic transport through a tube to the evaluation module is calculated by a PBM. As aggregation is assumed not to take place based on the experimental work of De Leersnyder et al. [14], only the last two terms in the right hand side of Equation (8) need to be used. The same breakage kernel as in the kneading zone of the TSWG was found to apply, i.e., breakage rate $S(\varepsilon)$ from Equation (10) and breakage fragment distribution $b(x,\varepsilon)$ from Equation (11). Parameter S_0 from from Equation (10) is linearly related to the remaining moisture content after drying.

Finally, the six cells are simulated in parallel with the *filling–drying–emptying* sequences according to the operation in the actual process. A cell is idle until the predecessor cell has been filled, at which point the current cell enters the filling stage. Hereafter, the remainder of the drying time is completed in the drying stage, where the mass flow rate at time t ($MFR_{TSWG}(t)$) in that cell is zero. In the final emptying stage, the mass transfer rates in Equations (13) and (15) are set to zero, and the change in

PSD is calculated by the PBM model based on the residual moisture content of the material. These material property values are thus those perceived at the inlet of the evaluation module.

2.1.5. Mill

In the wet granulation continuous manufacturing route, comilling is used to break the granulated product through collisions from a rotating impeller and walls. Granules that are broken to the required size exit the comill through a screen. In this work, the comill model published in Metta et al. [16] is used. Briefly, the mill model is a hybrid model that includes a PBM approach and a partial least squares (PLS) approach. Trajectories of change in mass of particles of various sizes over time is predicted through the PBM as shown in Equation (20) where $M(w,t)$ represents the mass of particles of volume w at time t, R_{form} and R_{dep} represent the rates of formation and depletion of particles respectively. $\dot{M}_{in}$ and $\dot{M}_{out}$ are the mass flow rates of particles entering and exiting the mill respectively. The rate of depletion R_{dep} is defined in the model (Equation (21)) using a breakage kernel $K(w)$, which represents the probability that a particle of volume w undergoes breakage. A classification kernel as given in Equation (22) is used in this work, where v_{imp} is the impeller speed, $v_{imp,min}$ is the minimum impeller speed and the parameter β is calibrated using data from experiments. The rate of formation R_{form} as shown in Equation (23) uses the breakage kernel and a breakage distribution function. The breakage distribution function $b(w,u)$ represents the distribution of daughter particles formed when a particle of volume w undergoes breakage. The Hill–Ng distribution function given in Equation (24) is used in this work, where the parameters p, q are estimated using experimental data. The mass flow rate out of the mill $\dot{M}_{out}(w,t)$ as given in Equation (25) is modeled using the feed particle size distribution (d_{in}) and a parameter $\Delta = d_{screen}\delta$, where d_{screen} is the screen size and δ is referred to as critical screen size ratio. A linear model is used to define the function f_d. The parameter δ is formulated as given in Equation (26) which represents the phenomenon of reduced apparent screen size available for a particle to exit the mill as impeller speed increases.

$$\frac{dM(w,t)}{dt} = R_{form}(w,t) - R_{dep}(w,t) + \dot{M}_{in}(w,t) - \dot{M}_{out}(w,t) \tag{20}$$

$$R_{dep}(w,t) = K(w)M(w,t) \tag{21}$$

$$K(w) = \begin{cases} \beta(\frac{v_{imp}}{v_{imp,min}})^2(\frac{w}{w_{ref}}) & \text{if } w \geq w(24) \\ 0 & \text{else} \end{cases} \tag{22}$$

$$R_{form}(w,t) = \int_{w}^{\infty} K(u)M(u,t)b(u,w)du \tag{23}$$

$$b(w,u) = \frac{p\frac{u}{w}^{q-1}(1-\frac{u}{w})^{r-1}}{wB(q,r)} \tag{24}$$

$$\dot{M}_{out}(w,t) = (R_{form}(w,t) - R_{dep}(w,t) + \gamma d_{in}(w,t))(1-f_d) \tag{25}$$

$$\delta = \epsilon(\frac{v_{imp,min}}{v_{imp}})^{\alpha} \tag{26}$$

Impeller speed showed little effect on the milled product when the comill feed is obtained from the fluid bed dryer because of breakage that occurs during transport to and from the fluid bed dryer in the horizontal ConsiGma^{TM}-25 configuration. Hence, v_{imp} is considered equal to $v_{imp,min}$ in the breakage kernel given in Equation (22).

The PBM is thus used to predict milled granule size distribution. The PLS model is an empirical modeling approach used to predict the milled product bulk density and tapped density, using the granule size distribution and moisture content as inputs. To use the mill model in the flowsheet model, batches of material are added to the milling unit each time the FBD completes a drying cycle and initiates an emptying cycle. When a batch of material is added to the existing material in the mill, breakage occurs. Properties of the material exiting the mill are obtained through mass averaging over the milling period t_{mill}. This is illustrated in Equation (27) for the bulk density of milled granules and is calculated the same way for the tapped density, span, and true density of the material exiting the mill. The instantaneous bulk density $\rho_{bulk,PLS}$ is obtained from the PLS model and weighted by the total mass flow rate of material exiting the mill, $\dot{M}_{out,total}$, at time t.

$$\rho_{bulk,milled} = \frac{\int_t^{t+t_{mill}} \rho_{bulk,PLS}(x)\dot{M}_{out,total}(x)dx}{\int_t^{t+t_{mill}} \dot{M}_{out,total}(x)dx} \tag{27}$$

2.1.6. Tablet Press

The tablet press model consists of four submodels. Firstly, a residence time distribution (RTD) model is used to describe the propagation of material properties through the powder dosing valve and the tablet press feed frame into the tablet die. The other models work with the material properties modeled as present in the die. The weight model relates material densities to the mean weight of the tablets semi-mechanistically (Equations (28) and (29)). The tablet potency model is a first-principles model (Equation (31)), and the tablet mean hardness model harbors literature empirical correlations related to tablet hardness and tensile strength (Equations (32) to (37)).

Propagation of the material properties through the feed frame is modeled according to a series of a continuously stirred tank reactor (CSTR) and a plug flow RTD model, with respective delay times t_{cstr} and t_d related to the feed frame turret speed. The solution of the feed frame model is performed in the same way as the transfer line models explained in Section 2.1.8. The tablet mean weight model uses the tooling dimensions (cup volume V_{cup}, die surface A_{die}, and punch cup depth D_{cup}), the material densities at the die (true density ρ_{true}, bulk ρ_{bulk} and tapped ρ_{tapped} density), a fill density factor $p_{\rho_{fill}}$, and the fill depth d_{fill} to calculate the mean tablet weight $\overline{m}$, according to the following equations.

$$\rho_{fill} = \rho_{bulk} + p_{\rho_{fill}} \left(\rho_{tapped} - \rho_{bulk}\right) \tag{28}$$

$$\overline{m} = (V_{cup} + A_{die}d_{fill})\rho_{fill} \tag{29}$$

The volume of the solids in the die V_{solid} is then:

$$V_{solid} = \overline{m}/\rho_{true} \tag{30}$$

The potency P simply follows from the API concentration C_{API} and $\overline{m}$:

$$P = \overline{m}\, C_{API} \tag{31}$$

The hardness model in the end uses V_{solid} from the tablet mean weight model, the main compression height process setting MCH and the tablet dimensions (width W, thickness T, and upper punch penetration upp) to calculate the tablet hardness through estimation of the tablet tensile strength. W and T are calculated according to:

$$W = MCH - upp \tag{32}$$

and

$$T = (MCH - upp) + 2D_{cup}, \tag{33}$$

resulting in a tablet volume V_{tablet} and relative density ρ_{rel} through:

$$V_{tablet} = (WA_{die}) + 2V_{cup} \tag{34}$$

and

$$\rho_{rel} = \frac{V_{solid}}{V_{tablet}}. \tag{35}$$

In order to relate the experimentally obtained hardness N values to the material properties for the biconvex tablets, a tensile strength σ_T normalization needs to be applied w.r.t. the tablet dimensions [20].

$$\sigma_T = \frac{10N}{\pi \mathrm{D}^2}\left(2.84\frac{\mathrm{t}}{\mathrm{D}} - 0.126\frac{\mathrm{t}}{\mathrm{W}} + 3.15\frac{\mathrm{W}}{\mathrm{D}} + 0.01\right)^{-1} \tag{36}$$

The tensile strength is linked to the critical density for tableting of the material ρ_c, and maximum achievable tensile strength σ_{max} [21]:

$$\sigma = \sigma_{\max}\left[\rho_c - \rho_{rel} - \ln\left(\frac{1-\rho_{rel}}{1-\rho_c}\right)\right]. \tag{37}$$

The constants of σ_{max} and ρ_c are obtained through empirical linear correlations with the relative size span R_{span} of the material and the moisture content X.

Finally, two modes of using the tablet press model are defined. In the *process setting* mode, tablet press process setting values (fill depth and MCH) are supplied to the model along with the material properties. Tablet properties such as the mean weight and hardness then result from the combination of the user defined process setting values with the incoming properties. These might however not match and result in unrealistic tablet properties as well as infeasible calculations, such as in the case where Equation (38) is not satisfied. In this mode, this is amended by the inclusion of a boolean variable that is only `True` when the input variables are such that condition Equation (38) is fulfilled.

$$\rho_c < \rho_{rel} < 1 \tag{38}$$

Another mean weight control (MWC) mode was created to allow for the tablet press model to calculate its process setting values in order to achieve a certain mean weight and hardness. These two variables thus serve as an input to the model in the flowsheet model, and the fill depth and MCH are calculated. This allows avoidance of the situation where user-specified tablet press process settings are not compatible with the simulated material properties governed by the process setting values upstream, possibly leading to a long simulation where no tablet weights or hardnesses could be calculated.

2.1.7. Overview

Once the individual unit operation models are developed and parameterized, the flowsheet model is built via connecting the inlet of a unit to the outlet of the preceding unit. Required information (material properties, operating conditions, etc.) are transferred from one unit to the succeeding unit as shown in Figure 1. The interaction between various modeling approaches in the individual unit operations is pictorially shown in Figure 2 to illustrate how empirical, semi-empirical, statistical, and mechanistic models interact to yield a flowsheet model. The system is solved using a backward differentiation solver with variable time steps, which is one of the two built-in solvers in gPROMS.

2.1.8. Intermediate Feeders, Blender, and Transfer Lines

The ConsiGma^{TM}-25 line includes an intermediate feeder for feeding powder blend from blender to the granulator (hereafter referred to as 'powder blend feeder'), and an intermediate feeder for feeding granulated material from the mill (hereafter referred to as 'granule feeder'). Thus, to achieve a more accurate representation of the ConsiGma^{TM}-25 line, transfer lines and the intermediate feeder and blender units are also included in the flowsheet model. A key difference in the implementation of the feeder model to the intermediate units is that the powder blend feeder and granule feeder do not have refilling units assigned to them, as these units receive material from their preceding units (powder blend feeder receives material from the blender and granule feeder receives material from the mill).

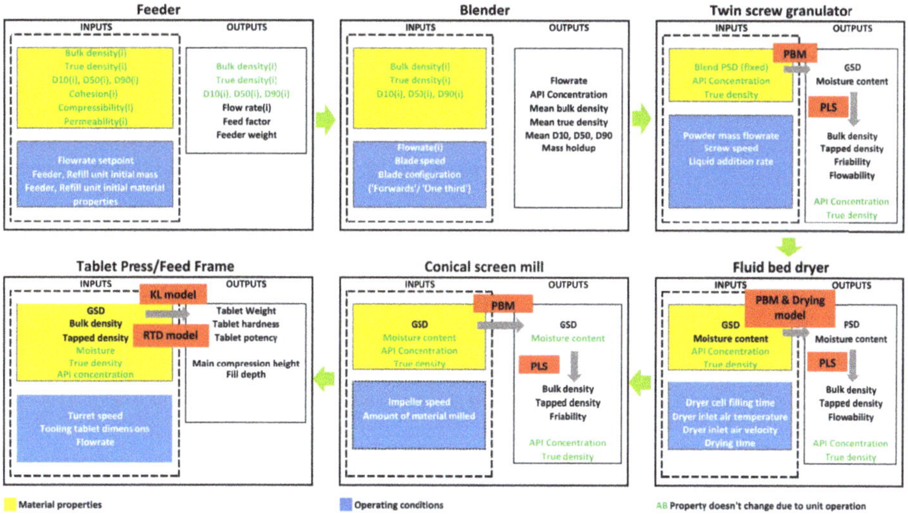

Figure 2. Schematic showing interaction between units and various modeling approaches.

In addition, a blending unit is also included to add lubricant to the granulated material from the mill. This intermediate blender will hereafter be referred to as 'granule blender'. The granule blender is modeled based on the axial dispersion equation as explained in Section 2.1.2. The blender is assumed to be filled with granulated material and the axial dispersion equation is used to transfer information regarding new or changing material properties from the upstream mill and the lubricant feeder. The true density of lubricated material from the granule blender is taken as weighted average of bulk densities of the milled granule product and lubricant.

In the dynamic flowsheet simulation, an accurate representation of the material properties at all times and at every location in the process would not be complete without accounting for material hold-up between unit operations. Hereto, transfer line models have been implemented to propagate variable values in between unit operation models. These models delay the propagated values according to a plug flow regime, assuming no back-mixing or axial dispersion is taking place during the transfer of materials between unit operations.

The plug flow propagation of these materials is often modeled applying a convolution of the inlet concentration profile at the modeled system, with the residence time distribution function of the material, as for instance described in [22]. This convolution requires information on the inlet concentration values over a range of time, which is not accessible for calculation in gPROMS. Therefore, this plug flow behavior is emulated using an axial dispersion model. A new simulation domain z is created to represent the normalized length of the transfer line under consideration, therefore this domain [0, L] is always equal to [0, 1]. Over this domain, the change in input signal S over time t at $z = 0$ is propagated over the domain as in Equation (39) where the plug flow time delay is represented as τ_{delay}. The input and output of the transfer lines are hence given by $S(t, 0)$ and $S(t, 1)$ respectively. The smaller ∂z is chosen, the smoother and more accurate the signal will propagate through the domain z, yet more computational burden is involved with this choice of more grid points. A value of $1/1000$ for ∂z has been found to give a good balance between smoothness and computational burden. Finally, it has been found that normalization of the delayed signal S drastically improves CPU time in the gPROMS solvers.

$$\tau_{delay}\frac{dS(t,z)}{dt} = \frac{\partial S(t,z)}{\partial z} \tag{39}$$

The above unit operation models are connected to develop the flowsheet model representing the ConsiGma^{TM}-25 line as shown in Figure 3. The flowsheet includes intermediate units and transfer lines as well. When the flowsheet model is simulated, it is important to have a clear understanding of the initial states of the model as it impacts the dynamic model state. For example, feeder hoppers could have various amounts of material at the start of the simulation. Similarly, the mill could start empty or have a certain amount of mass held up at the start of the simulation. Table 3 gives an overview of initial and dynamic states of the various units in the developed flowsheet model. The full flowsheet model thus developed can be used to simulate the continuous process as described in Section 3.1. Two case studies are provided in Section 3.2 where the full flowsheet model is used to understand the effect of step changes in process settings. The flowsheet model can also be used for advanced process analyses as described in the next sections.

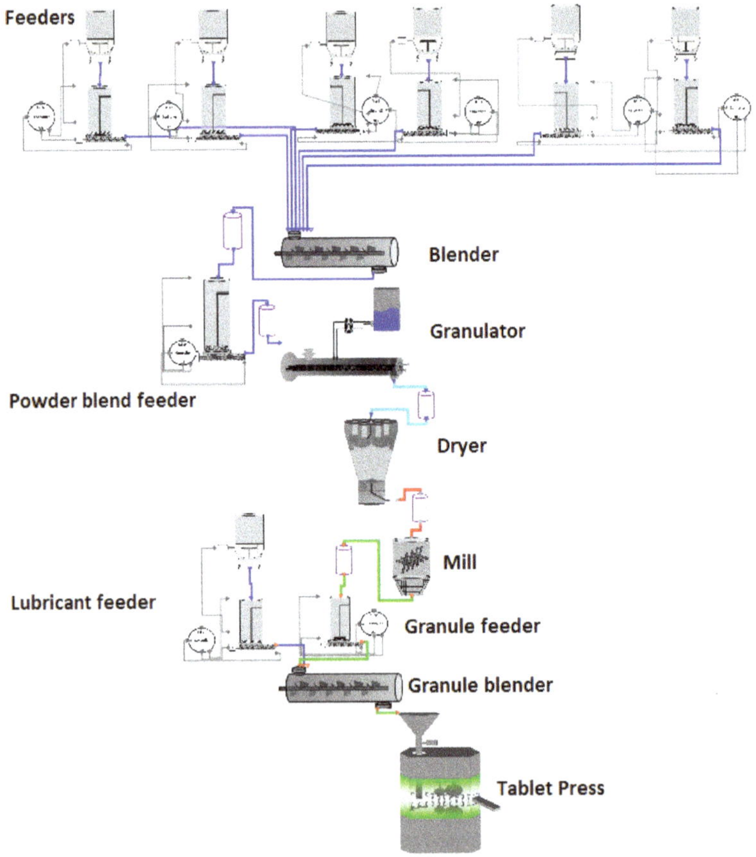

Figure 3. gPROMS Formulated Products schematic of the full flowsheet model developed.

Table 3. State of unit models used in flowsheet model development.

Unit	Initial State	Dynamic State
Feeders	Full	Refills when empty
Blender	Empty	Reaches steady state
Granulator	Empty	Output obtained when input is in studied range Plug flow delay added to instantaneous response
Dryer	Empty	Releases batches of material at the end of drying time
Mill	Empty	Releases material in semi-continuous mode
Tablet Press	Empty	Output obtained when input is in studied range Delay from RTD model in feed frame
Powder blend feeder	Full with powder blend	Continuous feed from blender
Granule feeder	Full with granules	Refill from mill
Granule blender	Always full	Delay from axial dispersion

2.2. *Scenario Analysis*

It is important to ensure that the flowsheet model developed successfully runs simulations at various process conditions. This affirms that the models are successfully integrated. In addition, it is important to verify that the process responses from these simulations are aligned with process knowledge. Scenario analysis provides a structured framework to achieve this, as several simulations at various combinations of process settings can be run in parallel and the resulting process responses can be analyzed. In this work, only process responses at the end of simulation, i.e., steady state tablet properties, are analyzed.

However, this exercise is computationally demanding as one simulation takes several hours to run. In this work, a more pragmatic approach is applied and the flowsheet model as given in Figure 3 is adapted in order to implement scenario analysis. Specifically, the intermediate units (powder blend feeder, granule feeder, granule blender) and the transfer lines are not considered. Since only the end of simulation responses are studied, the flowsheet model simplification is valid as the intermediate units and transfer lines do not affect the steady state process responses. The modified model lowers the computational expense and allows implementation of study required to affirm that the process responses from integration of unit operation models are meaningful. The full flowsheet model takes approximately 4 h to run a six-cell drying cycle, whereas the modified model only takes approximately 7 min for the tablet properties to reach steady state. Hence, the adapted model as shown in Figure 4 is used for the implementation of scenario analysis. In addition, the factors used for the study are also chosen judiciously in order to keep the total number of flowsheet model evaluations low. For example, factors such as blender blade speed, mill impeller speed are not considered. The blender blade speed influences the mass holdup in the blender, as explained in Section 2.1.2, but not the steady state flow rate, as this depends on the incoming feed flowrate. The mill impeller speed does not have an effect on granules obtained from a FBD [16]. Overall, the factors considered, their corresponding lower and upper bounds, and number of levels for each factor is listed in Table 4. In addition, the process responses that are recorded for each simulation run are also listed in Table 4. Three levels for flow rate setpoint, and four levels for LS ratio, granulator screw speed, dryer air temperature, and drying time, are chosen. In Section 3.3, an analysis of process responses from the simulations run is discussed in detail.

While the adapted flowsheet model is useful in implementing further analyses, it is also important to understand its limitations. The simplified flowsheet does not capture the effects of the intermediate units. For example, effects of refilling and propagation of disturbances from the intermediate units are ignored. In addition, a successful scenario analysis on the simplified model does not capture scenarios where the full flowsheet model fails due to the intermediate units. The simplified flowsheet does not support study of dynamic behavior of the line. Analyses such as dynamic sensitivity analysis or identification of dynamic feasible region cannot be accomplished.

Table 4. Factors and responses for scenario analysis and sensitivity analysis.

Unit	Factor	Bounds	Number of Levels	Response
Blender	flowrate setpoint, kg/h	[10, 20]	3	Mean Residence time Number of tanks
Granulator	LS ratio, kg/kg Screw speed, rpm	[0.08, 0.18] [500, 900]	4 4	PSD: d10, d50, d90 Moisture content
Dryer	Air temperature, deg C Drying time, s	[40, 60] [200, 1080]	4 4	PSD: d10, d50, d90 Moisture content
Mill				PSD: d10, d50, d90 Span Bulk density Tapped density
Tablet Press				Tablet hardness Tablet potency MCH Fill depth

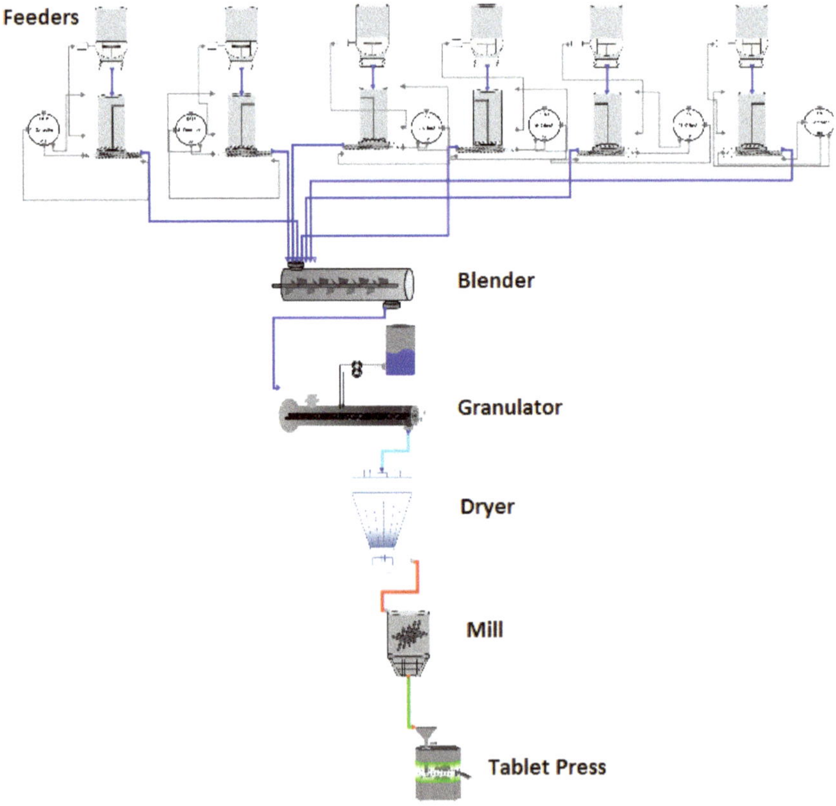

Figure 4. Schematic of the flowsheet model used for scenario analysis and sensitivity analysis.

2.3. *Sensitivity Analysis*

Sensitivity analysis is one of the key process systems engineering tools that can be used for quality risk assessment through identification of critical process parameters (CPPs). Sensitivity analysis is the investigation of how variability in the model inputs contributes to variations in model outputs [23]. It is an effective tool to rank and prioritize the process variables based on the effects they have on the output variables of interest. In the context of operation of a continuous manufacturing line, it helps identify the source of issues in meeting product quality or production demands. In the context of process model development, sensitivity analysis has been extensively used to identify the parameters that affect model outputs, thus helping focus experimental and model calibration efforts [23,24]. This helps researchers identify the areas where further model development needs to be focused on. Specifically for a flowsheet model where there is high number of input factors, sensitivity analysis can be used to reduce the number of input factors that need to be studied further. This helps in simplifying a high dimensional problem by filtering out the variables that have negligible effects on the outputs of interest. With this simplification, other tools can be applied for identification of design space of the process and its optimization [3].

Sensitivity analysis can be categorized into local and global methods. Local methods study the effect of input variables around a nominal point (or base case), whereas global methods study the effects over an entire input space. In this work, we focus on global sensitivity analysis as this is more relevant for pharmaceutical processes. For pharmaceutical processes, the input factors may include operating variables such as blender impeller speed, granulator screw speed, etc. The output variables of interest may include product properties such as tablet hardness, granule mean particle sizes, or process variables such as mill mass hold up, total flow rate, etc. There are various global sensitivity analysis methods available. The choice of method usually depends on the computational cost of evaluating the models, sampling budget available, and the detail of sensitivity information desired. In this work, Elementary effects method and Variance based sensitivity analysis methods are used, details of which are described in the next section. These methods are chosen as they are available in gPROMS FormulatedProducts with parallel computing capability.

2.3.1. Morris Method

Morris method also referred to as Elementary effects method is categorized under screening methods for sensitivity analysis. Screening methods are the most effective way to identify the most influential factors with relatively fewer samples [23]. Morris method is based on OAT (one-at-a-time) design where each of the input factors is varied and effects on the model outputs are studied.

For a model with k number of inputs, at a selected base point (x_1, x_2, ..., x_k), the elementary effect EE_i of the ith factor is given by Equation (40) where Δ_i is the step change in the ith input factor, Y represents the model output and $0 \leq \Delta_i \leq 1$. In order to represent the sensitivity information accurately, the sample points must be spread in the input space.

$$EE_i = \frac{Y(x_1, x_2, \ldots, x_i + \Delta_i, \ldots, x_k) - Y(x_1, x_2, \ldots, x_i, \ldots, x_k)}{\Delta_i} \tag{40}$$

Based on the calculation of EE_i, the sensitivity metrics as given in Equations (41)–(43) can be calculated, where r is the number of trajectories or radial base points for sampling. μ_i represents the average EE_i, σ_i^2 represents the variance and reflects non linearity or interactions in the ith input. μ_i* represents the average elementary effect using absolute EE_i to ensure the negative and positive effects do not cancel each other. It is suggested to look at all three metrics together to understand sensitivity information. Total sampling cost for this method is $r(k+1)$ where r can be less than 20 [25]. Hence, it is especially useful for models with a large number of input factors (factor of ten) or when the model is computationally expensive. In this work, the value of r is chosen as 20.

$$\mu_i = \frac{1}{r}\sum_{j=1}^{r} EE_i^j \tag{41}$$

$$\sigma_i^2 = \frac{1}{r-1}\sum_{j=1}^{r}(EE_i^j - \mu_i)^2 \tag{42}$$

$$\mu_i* = \frac{1}{r}\sum_{j=1}^{r} \mid EE_i^j \mid \tag{43}$$

After the metrics are obtained, input factors with large μ_i and/or μ_i*, σ_i^2 are considered to be significant. Practically, if the metric of an input factor is less than 10% of the largest value of this metric, the input factor is considered insignificant. While the method can be used to rank the factors, it does not quantify how much an input factor is more important than the other factors.

2.3.2. Variance Based Method

In this category of methods, the variance of the output is decomposed into several components including the individual inputs and the interactions between the inputs [26]. For an independent set of input factors, the variance $V(y)$ is expressed as given in Equation (44), where V_i is the variance term solely due to the input factor x_i, and $V_{i,j}$ is the variance term due to the interaction between the input factors x_i and x_j. Based on this variance decomposition, sensitivity measures can be defined as given in Equations (45)–(47).

$$V(y) = \sum_{i=1}^{k} V_i + \sum_{1\le i<j\le k} V_{i,j} + \ldots + V_{i,j,\ldots,k} \tag{44}$$

$$S_i = \frac{V_i}{V(y)} \tag{45}$$

$$S_{ij} = \frac{V_{i,j}}{V(y)} \tag{46}$$

$$S_{Ti} = \frac{V_i + \sum_{j\neq i} V_{i,j} + V_{1,2,..k}}{V(y)} = 1 - \frac{V_{\sim i}}{V(y)} \tag{47}$$

For the input x_i, S_i represents the 'first-order sensitivity index' whereas S_{ij} represents the 'second-order sensitivity index' which is the interaction effect of x_i and x_j on the process output. The metric S_{Ti} indicates the 'total sensitivity index', which accounts for the main effects as well as all the higher order interaction effects.

Specifically, this method uses Monte–Carlo techniques to compute the sensitivity indices as given in Equations (48) and (49) where $E(.)$ is the expected value and $X_{\sim i}$ represents all possible combinations of input factors with ith input factor X_i fixed. For this method, total number of samples required is $N(k+2)$ where N is recommended to be at least 500 [23]. In this work, the value of N is chosen as 500.

$$S_i = \frac{V_{X_i}(E_{X_{\sim i}}(Y \mid X_{\sim i}))}{V(Y)} \tag{48}$$

$$S_{Ti} = \frac{V_i + \sum_{j\neq i} V_{i,j} + V_{1,2,..k}}{V(y)} = 1 - \frac{V_{X_{\sim i}}(E_{X_i}(Y \mid X_{\sim i})}{V(y)} = \frac{E_{X_{\sim i}}(V_{X_i}(Y \mid X_{\sim i})}{V(y)} \tag{49}$$

For the variance based method, higher values of the metrics S_i and S_{Ti} indicate larger influence of the input factor. Also, S_i is always lower than or the same value as S_{Ti}. Hence, the difference between these metrics reflects interaction effects of the input factor with other factors. The adapted flowsheet model as explained in Section 2.2 is used for implementing the sensitivity analysis methods as well. In this work, details of factors and responses listed in Table 4 are also applicable for executing sensitivity analysis. In the next sections, results from simulating the flowsheet model, case studies to demonstrate dynamic simulation capabilities of the model and further analysis are presented and discussed.

3. Results and Discussion

3.1. Simulation Results

The flowsheet model as described in Section 2.1 is simulated using process settings as given in Table 5. The flow rate setpoints for the feeders are based on the formulation given in Table 1 and a total flow rate of 15 kg/h. The operating variables for other units were set within the ranges that were used to develop the individual unit operation models. The flowsheet model is simulated for 1500 s to complete a drying cycle using six dryer cells. Feeder levels in the seven component feeders (two API, one lubricant, and four excipient feeders) decrease until refill occurs at a fractional fill level of 0.1. The fill level in the powder blend feeder reduces until the blender starts feeding powder blend to it. The fill level in the granule feeder reduces until the mill starts feeding granules in a semi-continuous mode. Fill levels of all the component feeders and flow rate are shown in Figure 5a,b, respectively. Figure 5c shows fill levels and flow rates of the two intermediate feeders, i.e., powder blend and granule feeder.

Table 5. Values of process variables used for simulating the flowsheet model.

Unit	Process Variable	Units	Value
Feeders	API 1 flow rate	kg/h	11.337
	API 2 flow rate	kg/h	1.308
	Lubricant flow rate	kg/h	0.087
	Excipient A flow rate	kg/h	0.907
	Excipient B flow rate	kg/h	0.227
	Excipient C flow rate	kg/h	0.907
	Excipient D flow rate	kg/h	0.227
	Powder blend feeder flow rate	kg/h	14.913
	Granule feeder flow rate	kg/h	14.913
Blenders	Bladespeed	rpm	250
Granulator	Liquid-solid ratio	kg/kg	0.12
	Screw speed	rpm	500
Dryer	Air flow	m^3/h	360
	Air temperature	deg C	40
	Drying time	s	450
	Filling time	s	180
Tablet Press	Turret speed	rpm	29.8
	Mean weight	g	0.43

Powder blend flows continuously to the granulator. Wet granules start filling the first cell of the dryer and drying begins. After a filling time of 180 s, the second dryer cell starts filling. Dried granules in the first cell are emptied to the mill after a total drying time of 450 s. Thus, the cycle of filling, drying and emptying continues for the duration of simulation. Batches of dried granules fed to the mill are broken and leave the mill. The left axis in Figure 6a shows flow rate of the blend from the granulator. The right axis in Figure 6a shows mass of the batches of dried granules fed to the mill and corresponding change in holdup in the mill due to granules entering and leaving the mill. Figure 6b shows the evolution of moisture content of wet granules from the granulator and dry granules from the dryer as simulation progresses. As the simulation of the drying behavior requires a moisture content value at the beginning of its simulation, the results for the first dried batch are not representative because the wet granule moisture content at time $t = 0$ s equaled zero. This is the moisture content output of the FBD-model at time $t = 450$ s, which just indicates that the first dryer cell has emptied. This should be improved towards the future so that the FBD-model has a fully dynamic response towards its input (see Section 4).

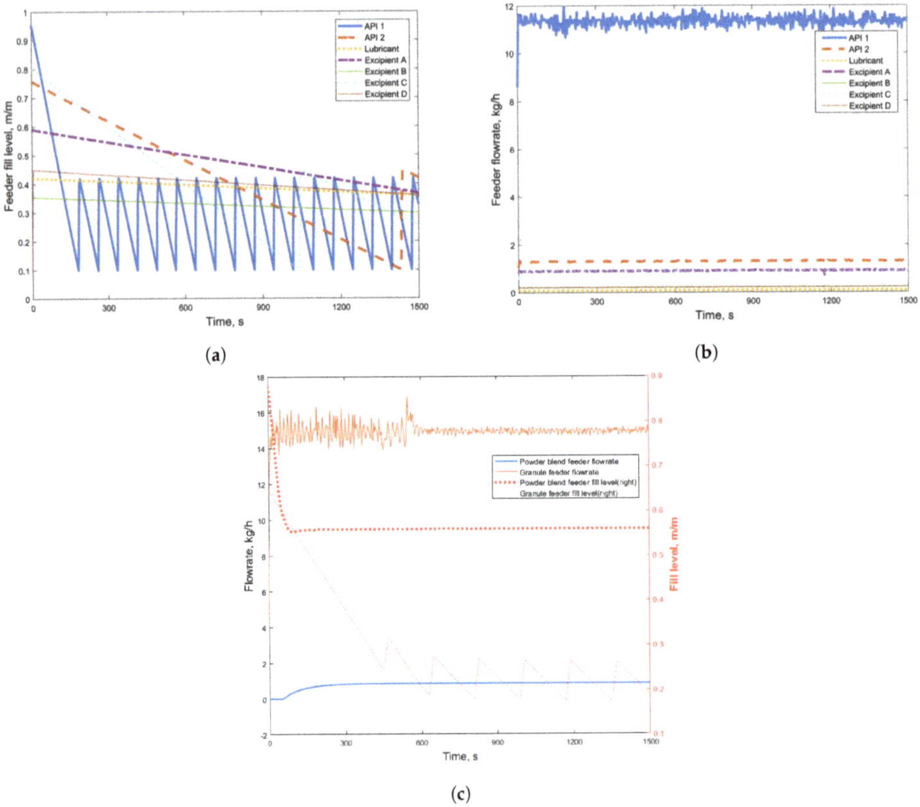

Figure 5. (**a**) Component feeder fill levels (**b**) Component feeder flow rates (**c**) Powder blend feeder and granule feeder fill levels and flow rates.

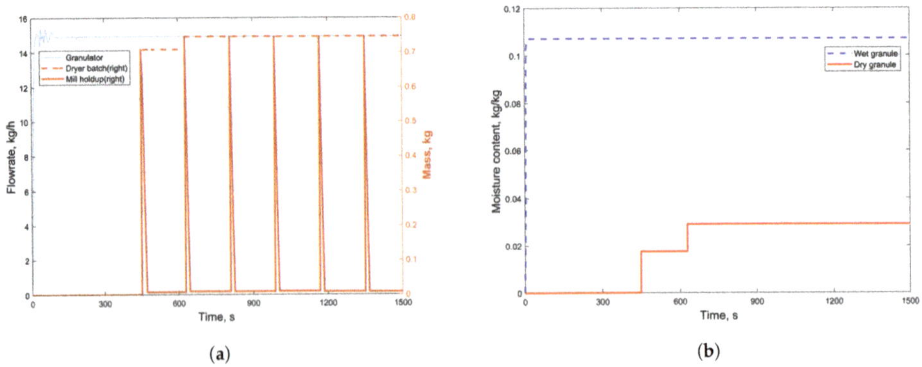

Figure 6. (**a**) Granulator flow rate, mass of granules from the dryer and holdup in mill (**b**) Moisture content of wet and dry granules.

As milled granules exit the mill, properties of the batch of milled granules are mass averaged and this information is propagated to the subsequent units. Figure 7a shows profiles of bulk density, tapped density, and true density of milled granules as the simulation progresses. Profiles of LOD, span, and API concentration of the milled product are shown in Figure 7b. It can be observed that the true density, LOD, and API concentration of milled product shows a step change around 600 s. This is because the intermediate powder blend feeder initially contains powder blend with a true density of 1291 kg/m^3, at the start of the simulation. This is eventually replaced when the powder blend of true density of 1344 kg/m^3, coming from the unit operations upstream, reaches that intermediate feeder in the simulation. Similarly, powder blend containing no API is replaced by a powder blend containing API 1 and API 2. This is reflected in the API concentration profile as shown in Figure 7b. In addition, LOD of the first batch of dried granules is lower (as shown in Figure 6b), which reflects in the milled granule LOD profile as well.

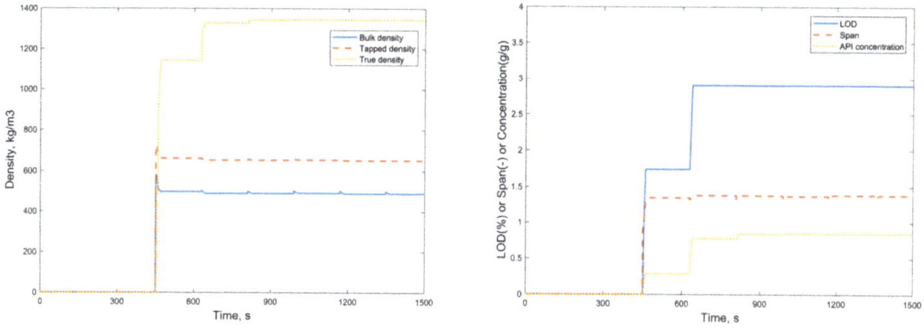

(**a**) (**b**)

Figure 7. Profiles of milled granules (**a**) bulk density, tapped density, and true density (**b**) LOD, span, and active pharmaceutical ingredient (API) concentration.

Milled granules fed to the granule feeder eventually replace granular material in it. Milled granules exiting the granule feeder are mixed in the granule blender with lubricant. Granules thus lubricated are sent to the feed frame, which is modeled as a PFR and CSTR in series. Propagation of properties (bulk density, tapped density, true density) in the granule feeder is shown in Figure 8a. The density profiles in this figure shows replacement of granules in the granule feeder with granules from the mill. Similarly, Figure 8b shows LOD, span, and API concentration profiles that simulate replacement of material existing in the granule feeder (2% LOD, span of 2, and 0.7 fractional API concentration) with milled granules from the upstream unit. Figure 9a shows density profiles of granules entering and leaving the feed frame. Similarly, Figure 9b shows profiles of LOD, span, and API concentration of granules entering and leaving the feed frame. The profiles changes seen in these figures is self explanatory based on the milled granule profiles (Figure 7a,b) and granule feeder profiles (Figure 8a,b).

The propagation of bulk density, tapped density, true density, API concentration, LOD, and span affect the profiles of tablet properties, namely, tablet hardness, weight, and potency. Figure 10a shows dynamic evolution of tablet properties as powder from component feeders replace material existing in the intermediate units (powder blend feeder and granule feeder). The tablet press hardness model was developed for material with bulk density greater than 300 kg/m^3. Hence, an initial tablet hardness of zero is shown in the hardness profile. In addition, since the tablet press is used in a mean weight control mode, main compression height and fill depth are adjusted as shown in Figure 10b in order to make tablets with a weight of 0.43 g.

3.2. Case Study

To clearly demonstrate the use of the flowsheet model for dynamic simulation purposes, two case studies are presented in this section. In both case studies, the full flowsheet model developed as explained in Section 2.1 is used. The initial simulation process variable settings are the same as explained in Section 3.1, Table 5. Later, in case study 1, a step change from 1.3 kg/h to 4 kg/h is made to the API 2 feeder flow rate setpoint at 200 s. In case study 2, a step change from 40 °C to 50 °C is made to the dryer air temperature at 455 s. A comparison between simulation results with and without the step changes implemented, and a demonstration of propagation of effects of the changes is discussed in the following sections.

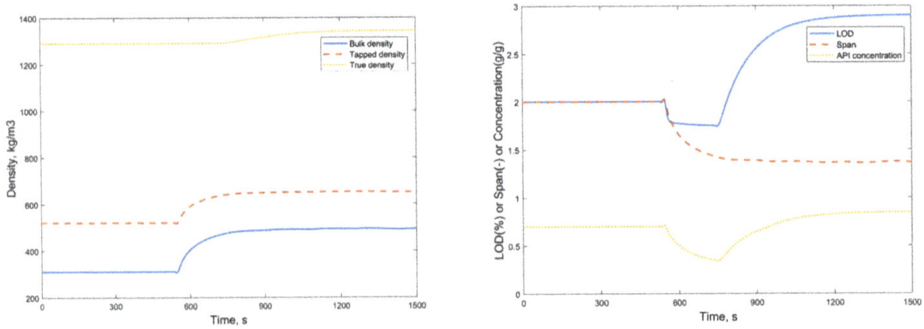

(**a**) (**b**)

Figure 8. Profiles of (**a**) bulk density, tapped density, and true density (**b**) LOD, span, and API concentration of granules from the granule feeder.

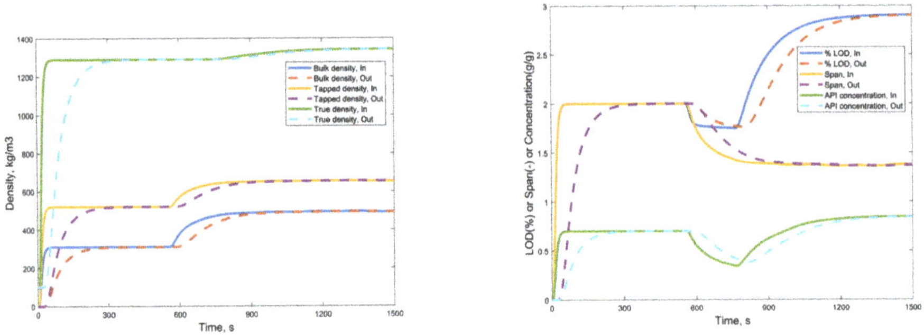

(**a**) (**b**)

Figure 9. Profiles of (**a**) bulk density, tapped density, and true density (**b**) LOD, span, and API concentration of granules entering and leaving the feed frame.

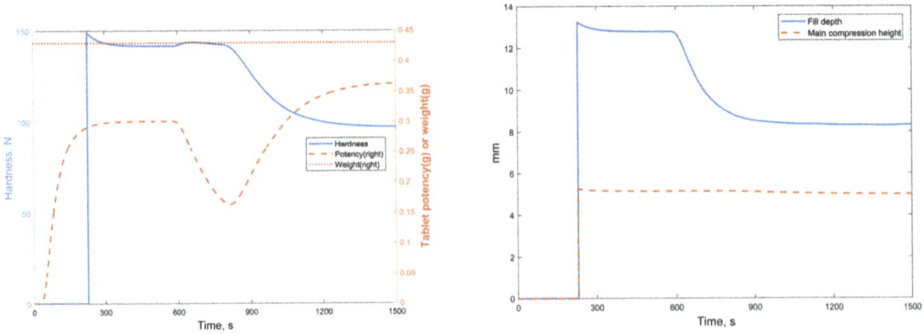

(**a**) (**b**)

Figure 10. Profiles of (**a**) tablet hardness, weight, and potency (**b**) tablet press fill depth and main compression height.

3.2.1. Case Study 1: Step Change in Feeder Flow Rate

A step change in API 2 feeder flow rate setpoint from an initial value of 1.3 kg/h to 4 kg/h is expected to lead to a change in API concentration (API 1 and API 2) in the powder blend, the granules, and subsequently potency of the tablets. Figure 11 shows the step change made in setpoint at 200 s leads to a change in API 2 feeder flow rate.

A change in fractional API concentration at the blender outlet from 0.85 to 0.87 is seen in Figure 12a as a result of the step change. Figure 12a also shows a change in API concentration of the powder blend leaving the powder blend feeder as initial material in the feeder (with no API) is eventually replaced by powder blend with API concentration of 0.87. A comparison with profiles from the simulation explained in Section 3.1 is also shown in this. Since the change is made at 200 s, the first cell in the dryer is filled (dryer filling time = 180 s) with granules from powder blend already present in the powder blend feeder. Hence, profiles of API concentration from the outlet of the granulator and dryer from the simulation explained in Section 3.1 and this case study are the same until the first dryer cell is emptied. This is shown in Figure 12b. Similarly, Figure 12c shows the propagation of change in API concentration at the outlet of respectively the mill and the granule feeder. As a result of this, a change in API concentration profiles at the outlet of granule blender and the feed frame are shown in Figure 12d. Finally, due to the step change in the amount of API 2 in the feed components an eventual deviation in the potency of tablets from 0.364 g to 0.374 g is shown in Figure 13. Thus, the case study demonstrates the use of the flowsheet model developed to track the effects of disturbances in upstream units on the final product quality.

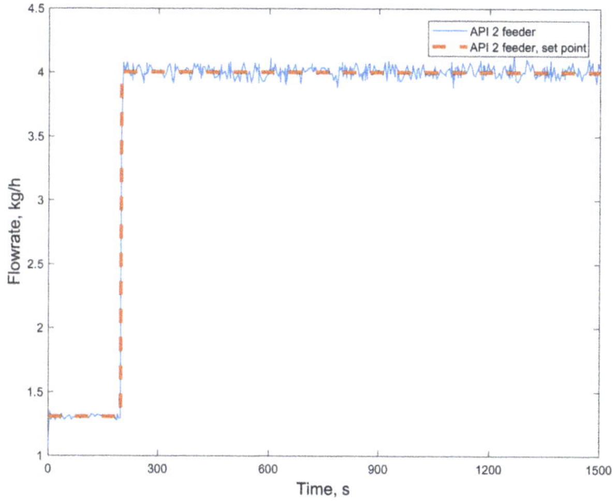

Figure 11. Step change in API 2 feeder flow rate setpoint showing an effect on the feeder flow rate.

3.2.2. Case Study 2: Step Change in Dryer Air Temperature

In case study 2, a step change from 40 °C to 50 °C is made to the dryer air temperature. The step change is made at 455 s as shown in Figure 14 (right axis). At 455 s, filling, drying, and emptying of the first cell are completed. In the second dryer cell, filling is completed and drying is in progress. In the third dryer cell, filling, and drying are in progress. Profiles of dried granule moisture content from the simulation explained in Section 3.1, along with this case study, are also plotted in Figure 14.

It can be observed that the dried granule moisture content from the first dryer cell is same in both cases as the step change occurs after emptying the first cell. In the second cell, dried granule moisture content is lower in the case where the step change is imposed. This is because granules in this cell are exposed to a higher air temperature for a duration of 175 s ($180 \times 1 + 450 - 455$), which leads to a lower granule moisture content. In the third cell, dried granule moisture content is further lowered as drying at 50 °C occurs for a longer duration of 355 s (180 × 2 + 450 − 455). In the fourth cell, the moisture content is further lowered to a steady state value as all the granules are dried at 50 °C.

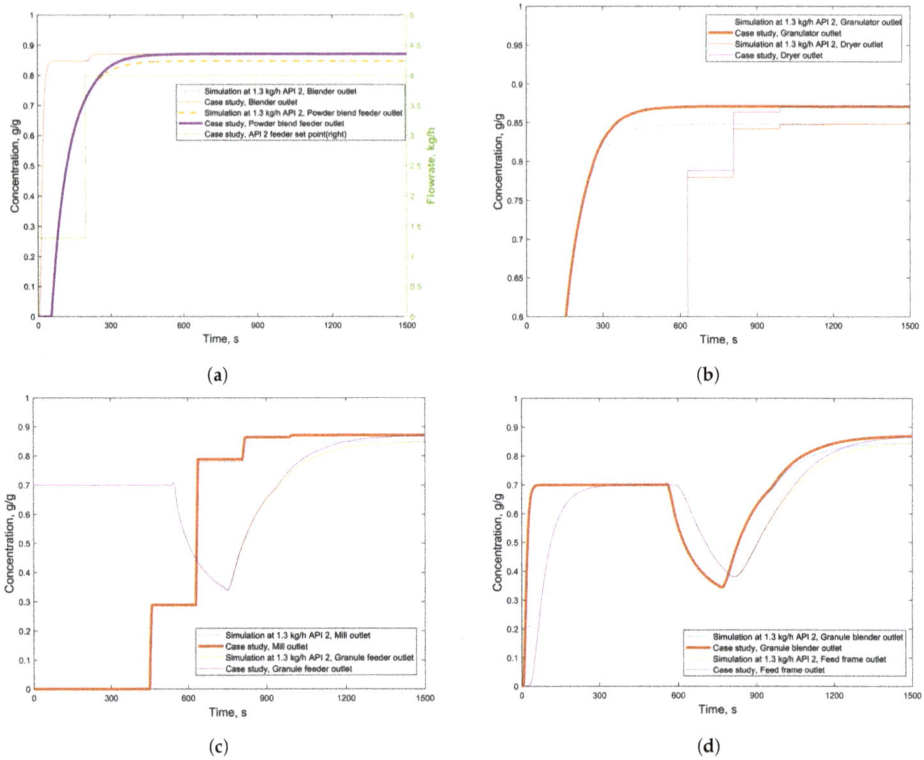

Figure 12. Comparison of API concentration profiles from simulations with fixed and step change in API 2 feeder flow rate setpoint for (**a**) blender and powder blend feeder (**b**) granulator and dryer (**c**) mill and granule feeder (**d**) granule blender and feed frame.

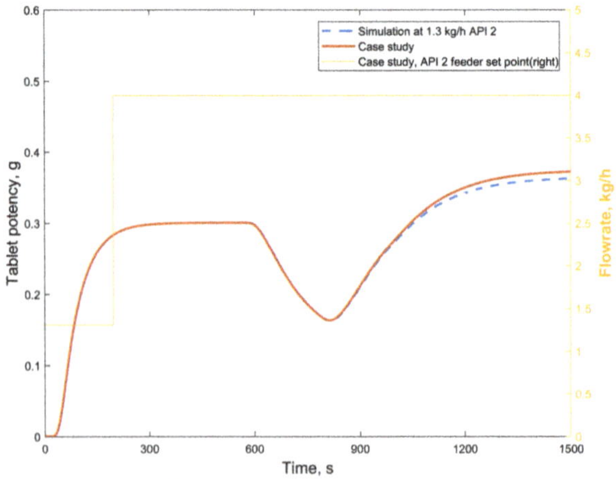

Figure 13. Comparison of tablet potency profiles from simulations with fixed and step change in API 2 feeder flow rate.

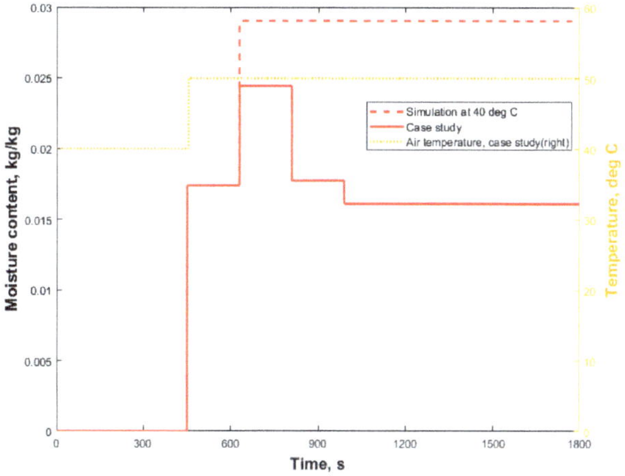

Figure 14. Comparison of dried granule moisture content profiles from simulations with fixed and step change in dryer air temperature.

Figure 15 shows propagation of LOD for both cases in the feed frame, which is the result of profile changes in the dryer and subsequent mill, feeder and blender units. It can be seen that the profiles follow the same path until about 750 s, after which a lower steady state value is reached for the case where step change to a higher air temperature occurs. The effect of difference in LOD profiles is also reflected in the tablet hardness as shown in Figure 16. Thus, implementation of the flowsheet model allows analysis of effects of such dynamic changes made to an upstream process variable on the final product quality of interest.

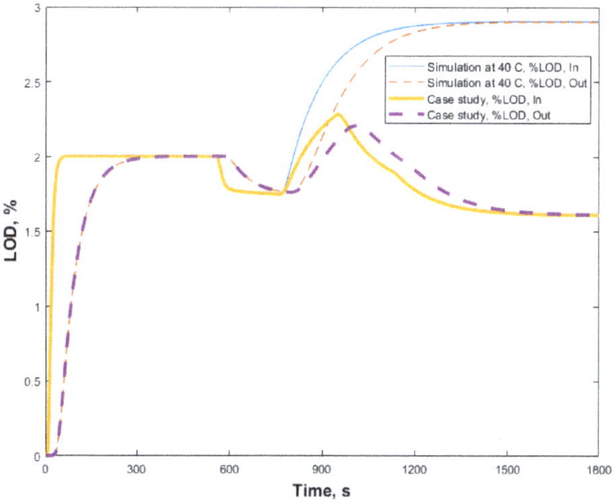

Figure 15. Comparison of moisture content of granules entering and leaving the feed frame from simulations with fixed and step change in dryer air temperature.

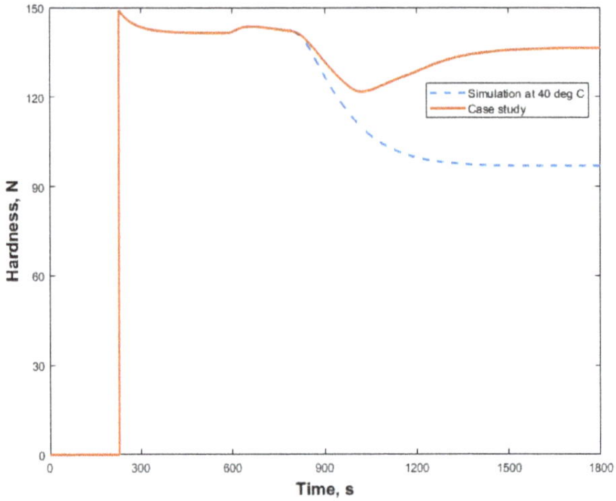

Figure 16. Comparison of tablet hardness profiles from simulations with fixed and step change in dryer air temperature.

3.3. Scenario Analysis Results

The adapted flowsheet model as shown in Figure 4 was used to implement a scenario analysis as explained in Section 2.2. Scenario analysis entails running the flowsheet model at various process setting values. It serves as a useful step before running a more computationally demanding sensitivity analysis. The flowsheet model is developed and errors are typically debugged at fixed values of process settings. However, this masks errors that may occur at other process setting values and does not provide confidence that the flowsheet model can run seamlessly. Debugging model errors at this stage before running more expensive analyses such as variance based sensitivity analysis serves as an effective modeling practice. Total flow rate, LS ratio, granulator screw speed, dryer air temperature, and drying time as tabulated in Table 4 are the input factors considered for scenario analysis. A total of 768 simulations from three levels of total flow rate, four levels each for LS ratio, granulator screw speed, dryer air temperature, and drying time ($3 \times 4 \times 4 \times 4 \times 4 = 768$) were successfully run. Process responses from blender, granulator, mill, dryer, and tablet press models as given Table 4 were recorded at the end of each simulation. For brevity, only few process responses are discussed in this section.

Specifically, wet granule d50, dry granule d50, dry granule LOD, and tablet hardness are plotted and discussed. Since, this is a multivariate analysis (five variables) plotting and analyzing responses from simultaneous change in all the variables is not possible. Hence, a matrix of plots as shown in Figure 17 are used. The matrix consists of 10 plots, each of which shows process response plots from varying two distinct factors with the three other factors fixed at baseline values. Figure 17a can be used to visualize and understand the effect of the five variables on wet granule d50. It can be observed that wet granule d50 increases with LS ratio and screw speed, and decreases with flow rate. This is in accordance with the experimental data used for granulator model development [13]. Figure 17b,c can be used to understand effects on dry granule size and moisture content respectively. Figure 17c shows that moisture content decreases with drying time which is an expected phenomenon. Figure 17b shows that dry granule size increases with LS ratio, screw speed, drying time, and air temperature, and decreases with flow rate. This is in accordance with experiments used for dryer model development [15]. Drying is expected to increase granule strength and lower breakage rate, which leads to a larger size. Hence, increase in drying time and air temperature increases granule size. The effect of flow rate, LS ratio, and screw speed on the dry granule size is due to the propagation of effects of these variables on the size of the wet granule feed to the dryer. Similarly, the effect of changes in the variables on tablet hardness is plotted in Figure 17d. It can be observed that low drying time leads to tablets with very low hardness. In other words, granules with high moisture content cannot be used to make tablets. This is in accordance with process knowledge as gained from experiments. It is also worth noting that development of a flowsheet model has enabled study of the effect of change in a process variable in an upstream unit on the final product quality.

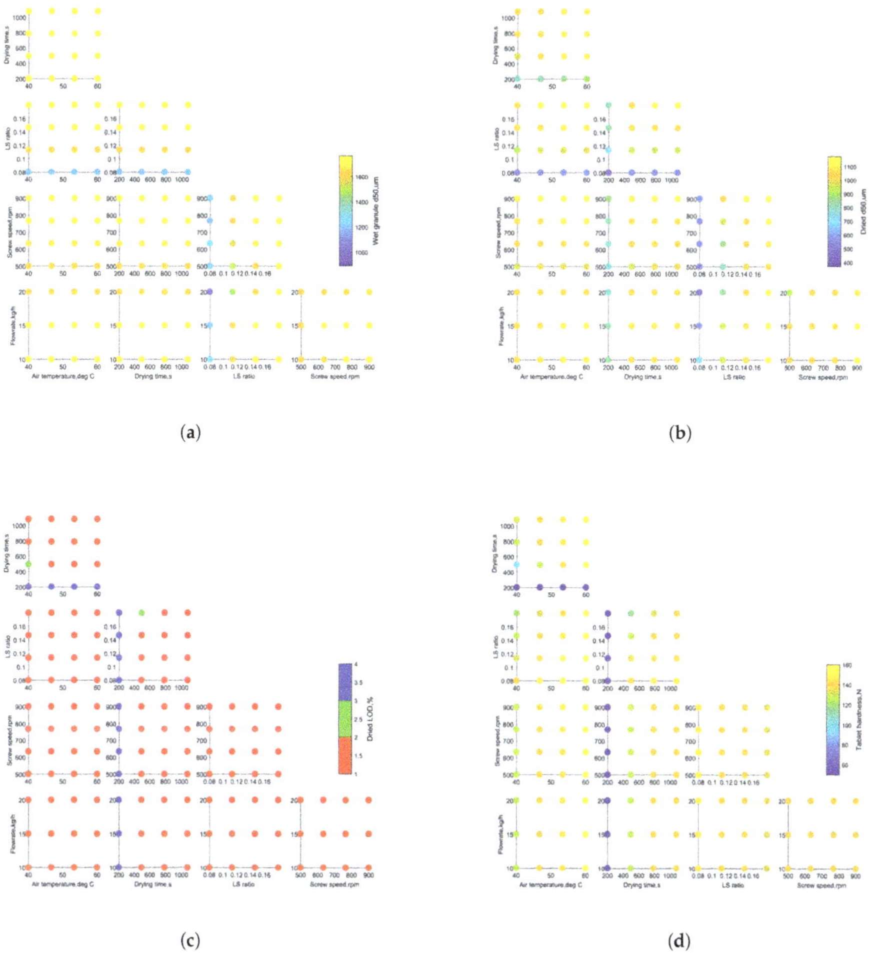

Figure 17. Scenario analysis plots for (**a**) wet granule d50 (**b**) dry granule d50 (**c**) dry granule LOD (**d**) tablet hardness.

3.4. Sensitivity Analysis Results

Sensitivity analysis was conducted using the adapted flowsheet model as explained in Section 2.2 and using five input factors, flow rate, LS ratio, granulator screw speed, dryer air temperature, and drying time. The effect of these input factors on 20 model responses are studied. The list of factors and responses, ranges for the factors are tabulated in Table 4. The input factors are considered to vary uniformly within these ranges. Morris analysis is implemented using 120 samples (= 20 × (5 + 1)). The results of the analysis are shown in Appendix A.1 that lists the metrics μ, $\mu*$, and σ. From the metrics, we observe that LS ratio influences wet granule moisture content, which is expected. Wet granule size is influenced by all three granulator process variables. This finding conforms with experiments that show effect of these variables on granulation rate [12]. For the dryer model outputs, flow rate and granulator screw speed do not show influence on dry granule moisture content and all five input factors show influence on dry granule size and dry granule properties (bulk density, tapped density, and angle of repose). For the mill model outputs, LS ratio shows the most influence on milled granule size, bulk, and tapped density. This is due to the effect of LS ratio on the size distribution of feed to the mill. The tablet press variables, main compression height and fill depth, are shown to be most influenced by air temperature, drying time and LS ratio. This is due to the effect of all these factors on the granule moisture content as granules with LOD higher than 3% cannot be used to make tablets. Thus, these factors also show an effect on tablet hardness.

From Morris analysis, dryer air temperature, drying time, and LS ratio are identified as the significant factors that influence tablet properties. Variance based analysis as explained in Section 2.3.2 can be applied to this subset of factors to obtain a quantitative understanding of their influence. However, for this work, variance-based sensitivity analysis is performed using all five factors and compared to results obtained from Morris method. The analysis was implemented using 3500 (= $500 \times (5 + 2)$) samples . First order sensitivity indices (S_i), as well as total sensitivity indices (S_{Ti}), were computed to quantify effects of the five input factors on 20 output responses. For brevity, the indices are tabulated in Appendix A.2. Here, only the total sensitivity indices, S_{Ti} are pictorially represented in Figure 18 as the first order effects S_i are close to S_{Ti} for responses from granulator, dryer, mill, and blender units. The responses from tablet press show interactions for the factors air temperature, drying time, and LS ratio. Generally speaking, findings from variance based sensitivity analysis agree with the findings from Morris analysis. The variance based method identified LS ratio as a significant factor for granulator. All five factors were identified as significant for dryer. LS ratio showed the greatest influence for the mill. All of these findings align with conclusions obtained from implementing Morris analysis. For the tablet press, air temperature, drying time, and LS ratio were identified as significant factors that influence tablet hardness, main compression height, and fill depth. In addition, all five factors were identified as significant for tablet potency. The influence of these factors on tablet potency was not observed from Morris analysis. This is because, in Morris analysis, the metrics are not dimensionless. For example, the metrics have a unit of g for tablet potency. Hence, any factor that shows an effect less than 0.01 g was identified as insignificant for Morris analysis. However, in variance based analysis, the metrics are dimensionless and the effect of various factors on the responses are normalized. A potency difference in the order of 1×10^{-6} g is also accurately identified in the sensitivity indices. Overall, both Morris and Variance-based methods serve in identifying critical factors. While, the Morris method requires fewer samples and allows ranking the factors by the order of influence, it does not provide quantitative information on relative effects of the factors. On the other hand, variance based analysis requires much higher number of samples but can provide detailed and quantitative information on the relative effects of the factors.

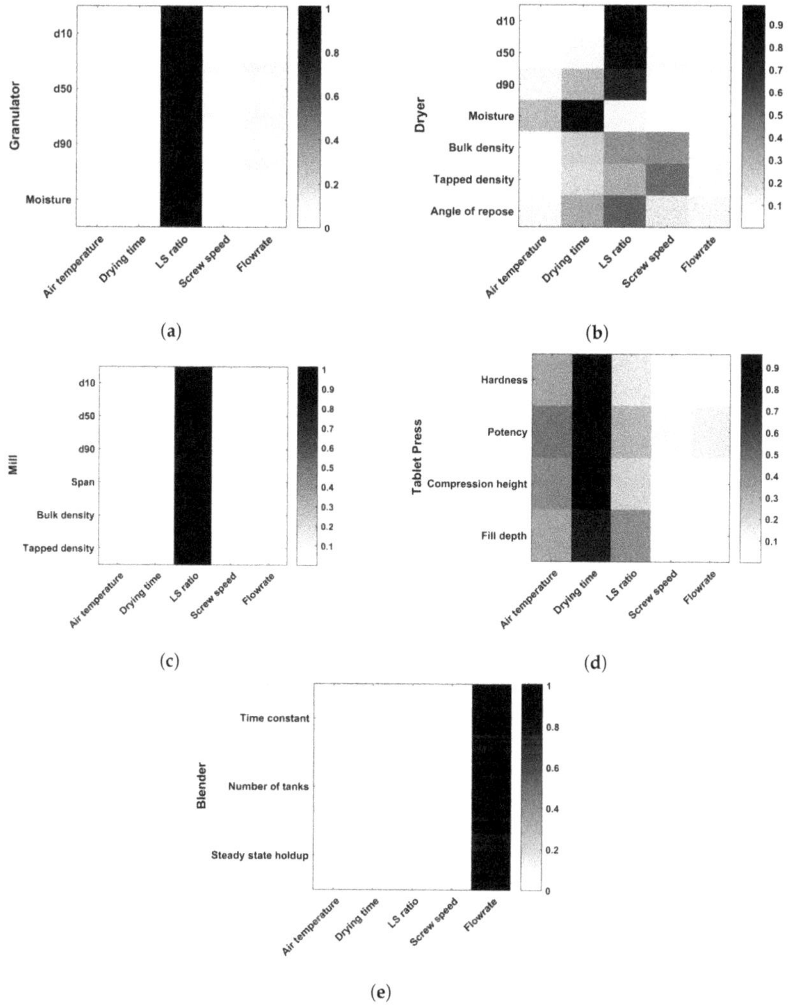

Figure 18. Total sensitivity index (S_{Ti}) plots for (**a**) Granulator (**b**) Dryer (**c**) Mill (**d**) Tablet press (**e**) Blender.

4. Conclusions and Future Direction

In this work, a flowsheet model that approximates the ConsiGma^{TM}-25 line for continuous tablet manufacturing through the wet granulation route is developed. The flowsheet model is based on models that are developed from experimental runs on units included in the continuous line using the same formulation and materials. For a complete virtual representation of the continuous line, models for intermediate units (powder blend feeder, granule feeder, and granule blender), as well as transfer lines, are also included in the flowsheet model. The developed model successfully demonstrates its ability to simulate the effect of changes in the process variables through case studies where step changes in API flow rate and dryer air temperature are implemented, and their effect on final tablet properties is understood. The robustness of the developed model is established by systematically running the flowsheet model at several combinations of process settings and analyzing the corresponding process responses. The model is also used to identify CPPs that affect intermediate and final product critical quality attributes (CQAs).

Throughout this article, several applications and capabilities of the developed flowsheet model have already been alluded to in the discussion. However, it is also worth noting some gaps in the developed model, which helps throw light on areas where future research efforts can be focused. For instance, the developed flowsheet model is computationally expensive. A simulation of about 1600 s takes approximately 4 h. While a simplified model is adapted in this work to implement steady state sensitivity analysis, it is not feasible to run dynamic sensitivity analysis using this model. Other areas of improvement include further development of the unit operation models. The blender model currently used predicts only 10%, 50%, and 90% percentile diameters for the powder blend. However, a much higher resolution PSD is required in the TSWG model. In addition, the TSWG model predicts only steady state PSD output based on process setting values. For the dryer model used in this work, drying behavior is based on input material properties at the start of the drying cycle of the cell. This could be further improved by incorporating dynamic modeling of drying behavior. For the mill model, the model parameters used currently are not a function of drying time in the fluid bed dryer. In addition, some of the submodels for unit operations use empirical relationships which are formulation specific. Another valuable verification of the model would be to check the mass balance over the entire system. The approximation of the axial dispersion models modeling the material flow propagation could thus be achieved. As research efforts continue on improving the unit operation models, the flowsheet model in its current state has already shown to be a useful tool for enhancing process understanding and enabling better decision making.

Overall, the developed flowsheet model is a prerequisite for identification of design space and optimization of the continuous line. Future research efforts should be focused on reducing computational expense of the model, as well as improving the capability of the unit models to capture dynamics and their applicability for other formulations suited for continuous solid oral dosage manufacturing.

Author Contributions: Conceptualization, all authors; Methodology, N.M. and M.G.; Software, N.M. and M.G.; Validation, N.M. and M.G.; Formal Analysis, N.M. and M.G.; Investigation, N.M. and M.G.; Resources, E.S., A.K., P.C., I.V.A., R.S., R.R., T.D.B., M.I. and I.N.; Data Curation, N.M. and M.G.; Writing—Original Draft Preparation, N.M. and M.G.; Writing—Review and Editing, E.S., A.K., P.C., R.R., T.D.B., M.I. and I.N.; Visualization, N.M. and M.G.; Supervision, E.S., A.K., P.C., I.V.A., R.S., R.R., T.D.B., M.I. and I.N.; Project Administration, E.S., A.K., P.C., I.V.A., R.S., R.R., T.D.B., M.I. and I.N.; Funding Acquisition, E.S., A.K., P.C., I.V.A.

Funding: This work is supported by a Consortium Agreement between Janssen Pharmaceutica, Ghent University and Rutgers University.

Acknowledgments: The authors gratefully acknowledge academic licenses from Process Systems Enterprise.

Conflicts of Interest: The authors declare no conflict of interest.

Acronyms

PBM	population balance model
GSD	granule size distribution
PSD	particle size distribution
TSWG	twin-screw wet granulator
FBD	fluid bed dryer
CPPs	critical process parameters
CQAs	critical quality attributes
API	active pharmaceutical ingredients
CSTR	continuously stirred tank reactor
RTD	residence time distribution
MWC	mean weight control
LIW	Loss-in-weight
PLS	partial least squares
QbD	Quality-by-Design
PSE	Process Systems Engineering

Nomenclature

Unit	Symbol	Description
General	t	time
Feeder	ff	Feed factor
	ff_{max}, ff_{min}, β	Model parameters
	ω	Screw speed
	$\dot{M}_{out}$	Mass flow rate, out
Blender	τ	Time constant
	M	Mass holdup
	M_{ss}	Steady state mass holdup
	$\dot{M}_{in}$	Mass flow rate, in
	$\dot{M}_{in}$	Mass flow rate, out
	τ_{ax}	Axial dispersion time constant
	C_{in}	API concentration, in
	C_{out}	API concentration, out
	Pe	Peclet number
	n_t	Number of tanks
Granulator	n	Number density
	x,ε	Particle volumes
	β	Collision frequency
	β_0	Aggregation efficiency
	b	Breakage fragment distribution
	β_0, R_1, R_2, top $_1$, top $_2$, δ_1, δ_2	Aggregation kernel parameters
	S	Breakage selection rate
	S_0	Breakage rate constant
	σ	Standard deviation of a Gaussian distribution
	μ	Mean of a Gaussian distribution
	f_{prim}	volume fraction of erosion in breakage
Dryer	$\dot{m}_v$	Mass transfer rate
	h_D	Mass transfer coefficient
	$\rho_{v,s}$	Partial vapor density over the droplet surface
	$\rho_{v,\infty}$	Partial vapor density in the ambient air
	A_d	Droplet surface area
	h_{fg}	Specific heat of evaporation
	$c_{p,w}$	Specific heat capacity liquid
	m_d	Droplet mass
	T_d	Uniform droplet temperature
	h	Heat transfer coefficient
	T_g	Drying gas temperature
	R_d	Droplet radius
	R_i	Wet radius
	R_p	Particle radius
	ϵ	Granule porosity
	β	empirical coefficient
	$D_{v,cr}$	Vapor diffusion coefficient
	M_w	Molecular weight liquid
	p_g	Pressure of the drying air
	$\Re$	Ideal gas constant
	$T_{cr,s}$	Temperature of solids at the granule surface
	$T_{wc,s}$	Temperature at particle gas-liquid interface
	$p_{v,\infty}$	Partial vapor pressure drying air
	$p_{v,i}$	Partial vapor pressure at gas-liquid interface
	$T_{p,s}$	Temperature of the particle solids
	X	Moisture content

	X_e	Equilibrium moisture content
	$\dot{X}$	Change in moisture content over time
	t_{fill}	Filling time
	t_{dry}	Drying time
	f	Size fraction
	$\overline{X}_f$	Average moisture content size fraction
	τ	Time
	n_τ	Dryer batch model parameter
	C_{API}	API concentration
	$MFR_{TSWG}(t)$	TSWG mass flow rate
Mill	M	Mass holdup
	w,u	Particle volumes
	R_{form}	Rate of formation
	R_{dep}	Rate of depletion
	$\dot{M}_{in}$	Mass flow rate, in
	$\dot{M}_{out}$	Mass flow rate, out
	K	Breakage kernel
	b	Breakage distribution function
	v_{imp}	Impeller speed
	$v_{imp,min}$	Minimum impeller speed
	d_{in}	Feed particle size distribution
	d_{screen}	Screen size
	Δ	Critical screen size
	$p, q, \beta, \epsilon, \alpha, \gamma, \delta$	Model parameters
	ρ_{bulk}	Bulk density
Tablet Press	t_{cstr}	Delay time CSTR
	t_d	Delay time plug flow
	V_{cup}	Tablet cup volume
	A_{die}	Tablet cup die surface
	D_{cup}	Tablet punch cup depth
	ρ_{true}	True density
	ρ_{bulk}	Bulk density
	ρ_{tapped}	Tapped density
	$p_{\rho_{fill}}$	Fill density factor
	d_{fill}	Fill depth
	$\overline{m}$	Mean tablet weight
	V_{solid}	Volume of solids in tablet die
	MCH	Main compression height
	W	Tablet width
	T	Tablet thickness
	upp	Upper punch penetration depth
	P	Tablet potency
	V_{tablet}	Tablet volume
	ρ_{rel}	Relative density
	N	Hardness
	σ_T	Tensile strength
	ρ_c	Critical density
	σ_{max}	Maximum tensile strength
	R_{span}	PSD size span
Intermediate units	S	Signal to axial disperion model
	z	Axial dispersion domain
	L	Transfer line length
	τ_{delay}	Plug flow time delay

Sensitivity analysis, Morris	EE_i	Elementary effect for factor i
	Y	Model output
	Δ_i	Step change in i_{th} input factor
	μ_i	Average elementary effect for factor i
	μ_i*	Average absolute elementary effect for factor i
	σ_i^2	Variance of elementary effects for factor i
Sensitivity analysis, Variance based	V_X	Variance of matrix X
	E_X	Expectede values for matrix X
	S_i	Sensitivity index for factor i
	S_{Ti}	Total sensitivity index for factor i

Appendix A

Appendix A.1. Morris Method Sensitivity Analysis

Granulator

Output →	d10, μm			d50, μm			d90, μm			Moisture, kg/kg		
Input ↓	μ	$\mu*$	σ	μ	$\mu*$	σ	μ	$\mu*$	σ	μ	$\mu*$	σ
Air temperature	0.11	0.21	0.30	0.52	0.87	1.29	0.98	1.77	2.59	0.00	0.00	0.00
Drying time	0.10	0.22	0.37	0.34	0.80	1.26	0.64	1.51	2.39	0.00	0.00	0.00
LS ratio	407.42	407.42	103.79	458.99	458.99	246.00	656.49	656.49	296.79	0.08	0.08	0.02
Screw speed	37.45	37.45	14.53	127.43	127.43	70.74	115.23	121.95	111.01	0.00	0.00	0.00
Flow rate	−35.49	38.10	28.06	−145.43	148.57	138.75	−148.87	154.33	136.70	0.00	0.00	0.00

Dryer

Output →	d10, μm			d50, μm			d90, μm			Moisture, %		
Input ↓	μ	$\mu*$	σ	μ	$\mu*$	σ	μ	$\mu*$	σ	μ	$\mu*$	σ
Air temperature	17.30	17.39	17.79	81.69	82.11	55.76	139.08	139.79	103.01	−2.01	2.01	1.64
Drying time	22.77	22.94	36.96	101.87	102.56	139.46	182.81	184.14	243.99	−2.77	2.77	3.50
LS ratio	273.71	273.71	81.03	410.60	410.60	206.75	307.38	326.94	200.84	1.50	1.50	1.93
Screw speed	18.03	18.03	7.37	121.74	121.74	53.72	86.86	86.86	48.09	0.00	0.00	0.00
Flow rate	−22.44	22.44	11.43	−109.86	115.99	99.38	−102.19	114.15	96.78	0.00	0.00	0.00

Output →	Bulk Density, Kg/m^3			Tapped Density, Kg/m^3			Angle of Repose, deg		
Input ↓	μ	$\mu*$	σ	μ	$\mu*$	σ	μ	$\mu*$	σ
Air temperature	−6.19	6.24	5.89	−4.75	4.78	4.53	0.79	0.80	0.62
Drying time	-8.30	8.33	12.12	−6.21	6.23	8.76	1.07	1.08	1.50
LS ratio	16.70	16.70	11.80	11.25	11.39	8.56	1.17	1.43	1.20
Screw speed	−14.25	14.25	5.73	−12.32	12.32	4.64	0.85	0.85	0.40
Flow rate	4.53	6.99	7.71	3.82	5.23	5.57	−0.63	0.80	0.76

Mill

Output →	d10, μm			d50, μm			d90, μm			Span		
Input ↓	μ	$\mu*$	σ	μ	$\mu*$	σ	μ	$\mu*$	σ	μ	$\mu*$	σ
Air temperature	4.81	4.81	5.06	7.94	7.96	5.06	8.76	8.80	6.68	−0.02	0.02	0.01
Drying time	6.50	6.56	10.44	9.45	9.52	12.07	11.21	11.27	13.85	−0.02	0.02	0.03
LS ratio	146.63	146.63	46.95	252.14	252.14	68.84	83.55	83.55	31.47	−0.96	0.96	0.29
Screw speed	3.38	3.68	3.56	3.16	9.13	10.91	16.16	16.53	10.26	0.01	0.03	0.03
Flow rate	−7.63	8.39	8.91	−8.83	13.65	12.13	−16.54	19.73	23.27	0.01	0.05	0.06

Output →	Bulk Density, Kg/m^3			Tapped Density, Kg/m^3		
Input ↓	μ	$\mu*$	σ	μ	$\mu*$	σ
Air temperature	4.93	4.93	3.24	4.09	4.09	2.54
Drying time	5.34	5.39	6.93	4.33	4.38	5.59
LS ratio	237.08	237.08	62.79	236.53	236.53	57.64
Screw speed	1.76	5.14	8.27	−16.09	17.69	14.84
Flow rate	−13.81	13.81	5.26	−8.02	12.52	14.14

Tablet press

Output →	Hardness, N			Potency, g			Compression Height, mm			Fill Depth, mm		
Input ↓	μ	$\mu*$	σ	μ	$\mu*$	σ	μ	$\mu*$	σ	μ	$\mu*$	σ
Air temperature	28.85	28.85	38.87	0.00	0.00	0.00	0.37	0.37	1.67	0.85	0.96	4.05
Drying time	84.89	84.89	100.13	0.00	0.00	0.00	2.62	2.62	3.71	4.57	4.63	6.83
LS ratio	−21.40	21.40	38.55	0.00	0.00	0.00	−0.37	0.37	1.67	−3.83	3.83	4.10
Screw speed	0.01	0.27	0.41	0.00	0.00	0.00	0.00	0.00	0.00	−0.02	0.14	0.24
Flow rate	0.06	0.49	0.66	0.00	0.00	0.00	0.00	0.00	0.00	0.22	0.22	0.21

Blender

Output →	Time Constant, s			Number of Tanks			Steady State Holdup, kg		
Input ↓	μ	$\mu*$	σ	μ	$\mu*$	σ	μ	$\mu*$	σ
Air temperature	0.00	0.00	0.00	0.00	0.00	0.00	0.00	0.00	0.00
Drying time	0.00	0.00	0.00	0.00	0.00	0.00	0.00	0.00	0.00
LS ratio	−0.00	0.00	0.00	0.00	0.00	0.00	0.00	0.00	0.00
Screw speed	0.00	0.00	0.00	0.00	0.00	0.00	0.00	0.00	0.00
Flow rate	−15.53	15.53	3.68	−0.15	0.15	0.07	0.03	0.03	0.01

Appendix A.2. Variance Based Sensitivity Analysis

Granulator

Output →	d10		d50		d90		Moisture	
Input ↓	S_i	S_{Ti}	S_i	S_{Ti}	S_i	S_{Ti}	S_i	S_{Ti}
Air temperature	0.00	0.00	0.00	0.00	0.00	0.00	0.00	0.00
Drying time	0.00	0.00	0.00	0.00	0.00	0.00	0.00	0.00
LS ratio	0.98	0.99	0.87	0.92	0.90	0.94	1.00	1.01
Screw speed	0.00	0.01	0.03	0.05	0.00	0.02	0.00	0.00
Flow rate	0.00	0.01	0.03	0.09	0.03	0.09	0.00	0.00

Dryer

Output →	d10		d50		d90		Moisture	
Input ↓	S_i	S_{Ti}	S_i	S_{Ti}	S_i	S_{Ti}	S_i	S_{Ti}
Air temperature	0.00	0.01	0.00	0.03	0.04	0.10	0.11	0.27
Drying time	0.01	0.02	0.05	0.08	0.17	0.27	0.64	0.85
LS ratio	0.97	0.98	0.78	0.84	0.58	0.70	0.00	0.08
Screw speed	0.00	0.01	0.04	0.07	0.01	0.05	0.00	0.00
Flow rate	0.00	0.01	0.01	0.06	0.00	0.06	0.00	0.00

Output →	**Bulk Density**		**Tapped Density**		**Angle of Repose**	
Input ↓	S_i	S_{Ti}	S_i	S_{Ti}	S_i	S_{Ti}
Air temperature	0.01	0.06	0.01	0.06	0.04	0.11
Drying time	0.12	0.20	0.11	0.18	0.21	0.32
LS ratio	0.24	0.38	0.19	0.30	0.38	0.53
Screw speed	0.40	0.41	0.49	0.50	0.11	0.15
Flow rate	0.00	0.08	0.00	0.07	0.00	0.11

Mill

Output →	**d10**		**d50**		**d90**		**Span**	
Input ↓	S_i	S_{Ti}	S_i	S_{Ti}	S_i	S_{Ti}	S_i	S_{Ti}
Air temperature	0.00	0.00	0.00	0.00	0.00	0.01	0.00	0.00
Drying time	0.00	0.00	0.00	0.00	0.00	0.02	0.00	0.00
LS ratio	0.99	1.00	0.99	1.01	0.91	0.96	0.99	1.01
Screw speed	0.00	0.00	0.00	0.00	0.02	0.04	0.00	0.00
Flow rate	0.00	0.01	0.00	0.00	0.00	0.03	0.00	0.00

Output →	**Bulk Density**		**Tapped Density**	
Input ↓	S_i	S_{Ti}	S_i	S_{Ti}
Air temperature	0.00	0.00	0.00	0.00
Drying time	0.00	0.00	0.00	0.00
LS ratio	1.00	1.01	0.99	1.00
Screw speed	0.00	0.00	0.00	0.01
Flow rate	0.00	0.00	0.00	0.00

Tablet press

Output →	**Hardness**		**Potency**		**Compression Height**		**Fill Depth**	
Input ↓	S_i	S_{Ti}	S_i	S_{Ti}	S_i	S_{Ti}	S_i	S_{Ti}
Air temperature	0.14	0.33	0.04	0.45	0.08	0.39	0.08	0.32
Drying time	0.58	0.88	0.52	0.85	0.51	0.96	0.40	0.72
LS ratio	0.03	0.14	0.01	0.28	0.02	0.21	0.26	0.40
Screw speed	0.00	0.00	0.00	0.07	0.00	0.00	0.00	0.00
Flow rate	0.00	0.00	0.02	0.13	0.00	0.00	0.00	0.00

Blender

Output →	**Time Constant**		**Number of Tanks**		**Steady State Holdup**	
Input ↓	S_i	S_{Ti}	S_i	S_{Ti}	S_i	S_{Ti}
Air temperature	0.00	0.00	0.00	0.00	0.00	0.00
Drying time	0.00	0.00	0.00	0.00	0.00	0.00
LS ratio	0.00	0.00	0.00	0.00	0.00	0.00
Screw speed	0.00	0.00	0.00	0.00	0.00	0.00
Flow rate	1.00	1.00	1.00	1.01	1.00	1.00

References

1. Escotet-Espinoza, M.S.; Singh, R.; Sen, M.; O'Connor, T.; Lee, S.; Chatterjee, S.; Ramachandran, R.; Ierapetritou, M.G.; Muzzio, F.J. Flowsheet Models Modernize Pharmaceutical Manufacturing Design and Risk Assessment. *Pharm. Technol.* **2015**, *39*, 34–42.
2. Gernaey, K.V.; Cervera-Padrell, A.E.; Woodley, J.M. A perspective on PSE in pharmaceutical process development and innovation. *Comput. Chem. Eng.* **2012**, *42*, 15–29. [CrossRef]

3. Wang, Z.; Escotet-Espinoza, M.S.; Ierapetritou, M. Process analysis and optimization of continuous pharmaceutical manufacturing using flowsheet models. *Comput. Chem. Eng.* **2017**, *107*, 77–91. [CrossRef]
4. Galbraith, S.C.; Huang, Z.; Cha, B.; Liu, H.; Hurley, S.; Flamm, M.H.; Meyer, R.J.; Yoon, S. Flowsheet Modeling of a Continuous Direct Compression Tableting Process at Production Scale. In Proceedings of the Foundations of Computer Aided Process Operations/Chemical Process Control, Tucson, AZ, USA, 8–12 January 2017.
5. Rogers, A.J.; Inamdar, C.; Ierapetritou, M.G. An Integrated Approach to Simulation of Pharmaceutical Processes for Solid Drug Manufacture. *Ind. Eng. Chem. Res.* **2014**, *53*, 5128–5147. [CrossRef]
6. García-Muñoz, S.; Butterbaugh, A.; Leavesley, I.; Manley, L.F.; Slade, D.; Bermingham, S. A flowsheet model for the development of a continuous process for pharmaceutical tablets: An industrial perspective. *AIChE J.* **2018**, *64*, 511–525. [CrossRef]
7. Park, S.Y.; Galbraith, S.C.; Liu, H.; Lee, H.; Cha, B.; Huang, Z.; O'Connor, T.; Lee, S.; Yoon, S. Prediction of critical quality attributes and optimization of continuous dry granulation process via flowsheet modeling and experimental validation. *Powder Technol.* **2018**, *330*, 461–470. [CrossRef]
8. Boukouvala, F.; Niotis, V.; Ramachandran, R.; Muzzio, F.J.; Ierapetritou, M.G. An integrated approach for dynamic flowsheet modeling and sensitivity analysis of a continuous tablet manufacturing process. *Comput. Chem. Eng.* **2012**, *42*, 30–47. [CrossRef]
9. Boukouvala, F.; Chaudhury, A.; Sen, M.; Zhou, R.; Mioduszewski, L.; Ierapetritou, M.G.; Ramachandran, R. Computer-Aided Flowsheet Simulation of a Pharmaceutical Tablet Manufacturing Process Incorporating Wet Granulation. *J. Pharm. Innov.* **2013**, *8*, 11–27. [CrossRef]
10. Boukouvala, F.; Ierapetritou, M.G. Surrogate-based optimization of expensive flowsheet modeling for continuous pharmaceutical manufacturing. *J. Pharm. Innov.* **2013**, *8*, 131–145. [CrossRef]
11. Rogers, A.; Ierapetritou, M. Challenges and opportunities in modeling pharmaceutical manufacturing processes. *Comput. Chem. Eng.* **2015**, *81*, 32–39. [CrossRef]
12. Verstraeten, M.; Van Hauwermeiren, D.; Lee, K.; Turnbull, N.; Wilsdon, D.; am Ende, M.; Doshi, P.; Vervaet, C.; Brouckaert, D.; Mortier, S.T.; et al. In-depth experimental analysis of pharmaceutical twin-screw wet granulation in view of detailed process understanding. *Int. J. Pharm.* **2017**, *529*, 678–693. [CrossRef] [PubMed]
13. Van Hauwermeiren, D.; Verstraeten, M.; Doshi, P.; am Ende, M.T.; Turnbull, N.; Lee, K.; De Beer, T.; Nopens, I. On the modelling of granule size distributions in twin-screw wet granulation: Calibration of a novel compartmental population balance model. *Powder Technol.* **2018**, *341*, 116–125. [CrossRef]
14. De Leersnyder, F.; Vanhoorne, V.; Bekaert, H.; Vercruysse, J.; Ghijs, M.; Bostijn, N.; Verstraeten, M.; Cappuyns, P.; Van Assche, I.; Vander Heyden, Y.; et al. Breakage and drying behaviour of granules in a continuous fluid bed dryer: Influence of process parameters and wet granule transfer. *Eur. J. Pharm. Sci.* **2018**, *115*, 223–232. [CrossRef] [PubMed]
15. Ghijs, M.; Schäfer, E.; Kumar, A.; Cappuyns, P.; Van Assche, I.; De Leersnyder, F.; Vanhoorne, V.; De Beer, T.; Nopens, I. Modeling of Semicontinuous Fluid Bed Drying of Pharmaceutical Granules With Respect to Granule Size. *J. Pharm. Sci.* **2019**. [CrossRef] [PubMed]
16. Metta, N.; Verstraeten, M.; Ghijs, M.; Kumar, A.; Schafer, E.; Singh, R.; De Beer, T.; Nopens, I.; Cappuyns, P.; Van Assche, I.; Ierapetritou, M.; Ramachandran, R. Model development and prediction of particle size distribution, density and friability of a comilling operation in a continuous pharmaceutical manufacturing process. *Int. J. Pharm.* **2018**, *549*, 271–282. [CrossRef] [PubMed]
17. Pantelides, C.C.; Nauta, M.; Matzopoulos, M.; Grove, H. Equation-Oriented Process Modelling Technology: Recent Advances & Current Perspectives. In Proceedings of the 5th Annual TRC-Idemitsu Work, Abu Dhabi, UAE, 15 February 2015.
18. Escotet Espinoza, M. Phenomenological and Residence Time Distribution Models for Unit Operations in a Continuous Pharmaceutical Manufacturing Process. Ph.D. Thesis, Rutgers, The State University of New Jersey, Brunswick, NJ, USA, 2018.
19. Abuaf, N.; Staub, F.W. Drying of Liquid–Solid Slurry Droplets. In *Drying '86*; Mujumdar, A.S., Ed.; Hemisphere: Washington, DC, USA, 1986; Volume 1, pp. 227–248.
20. Pitt, K.G.; Newton, J.M.; Richardson, R.; Stanley, P. The Material Tensile Strength of Convex-faced Aspirin Tablets. *J. Pharm. Pharmacol.* **1989**, *41*, 289–292. [CrossRef] [PubMed]

21. Kuentz, M.; Leuenberger, H. A new model for the hardness of a compacted particle system, applied to tablets of pharmaceutical polymers. *Powder Technol.* **2000**, *111*, 145–153. [CrossRef]
22. Engisch, W.; Muzzio, F. Using Residence Time Distributions (RTDs) to Address the Traceability of Raw Materials in Continuous Pharmaceutical Manufacturing. *J. Pharm. Innov.* **2016**, *11*, 64–81. [CrossRef] [PubMed]
23. Saltelli, A.; Ratto, M.; Andres, T.; Campolongo, F.; Cariboni, J.; Gatelli, D.; Saisana, M.; Tarantola, S. *Global Sensitivity Analysis: The Primer*; Wiley: Hoboken, NJ, USA, 2008.
24. Cryer, S.A.; Scherer, P.N. Observations and process parameter sensitivities in fluid-bed granulation. *AIChE J.* **2003**, *49*, 2802–2809. [CrossRef]
25. Iooss, B.; Lemaître, P. A Review on Global Sensitivity Analysis Methods. In *Uncertainty Management in Simulation-Optimization of Complex Systems: Algorithms and Applications*; Dellino, G., Meloni, C., Eds.; Springer: New York, NY, USA, 2015; pp. 101–122.
26. Helton, J.C.; Davis, F.J. Latin hypercube sampling and the propagation of uncertainty in analyses of complex systems. *Reliab. Eng. Syst. Saf.* **2003**, *81*, 23–69. [CrossRef]

processes

MDPI

Article

An Optimization-Based Framework to Define the Probabilistic Design Space of Pharmaceutical Processes with Model Uncertainty

Daniel Laky [1,†], Shu Xu [1,†], Jose S. Rodriguez [1], Shankar Vaidyaraman [2], Salvador García Muñoz [2] and Carl Laird [1,3,*]

1 Davidson School of Chemical Engineering, Purdue University, West Lafayette, IN 47907, USA; dlaky@purdue.edu (D.L.); richard041123@gmail.com (S.X.); rodri324@purdue.edu (J.S.R.)
2 Small Molecule Design and Development, Lilly Research Laboratories, Eli Lilly & Company, Indianapolis, IN 46285, USA; shankarraman_vaidyaraman@lilly.com (S.V.); sal.garcia@lilly.com (S.G.M.)
3 Sandia National Laboratories, Albuquerque, NM 87123, USA
* Correspondence: lairdc@purdue.edu
† Both authors contributed equally to this work.

Received: 29 December 2018; Accepted: 1 February 2019; Published: 14 February 2019

Abstract: To increase manufacturing flexibility and system understanding in pharmaceutical development, the FDA launched the quality by design (QbD) initiative. Within QbD, the *design space* is the multidimensional region (of the input variables and process parameters) where product quality is assured. Given the high cost of extensive experimentation, there is a need for computational methods to estimate the *probabilistic* design space that considers interactions between critical process parameters and critical quality attributes, as well as model uncertainty. In this paper we propose two algorithms that extend the flexibility test and flexibility index formulations to replace simulation-based analysis and identify the probabilistic design space more efficiently. The effectiveness and computational efficiency of these approaches is shown on a small example and an industrial case study.

Keywords: pharmaceutical processes; flexibility analysis; probabilistic design space; global optimization

1. Introduction

To increase manufacturing flexibility, process robustness, system understanding, and to prevent the shortage of critical medicines due to unreliable quality in pharmaceutical development and manufacturing, the FDA launched the quality by design (QbD) initiative [1]. Later, the concept of the design space was characterized as "the multidimensional combination and interaction of input variables (e.g., material attributes) and process parameters that have been demonstrated to provide assurance of quality" [2]. On one hand, the design space offers operational flexibility for industries to continuously improve performance as long as the combination of input variables and process parameters fall within the approved design space [3]; on the other hand, the design space provides regulatory agencies with a convenient tool to monitor the compliance of a pharmaceutical production process [4].

The design space is identified by the limits of acceptability of critical quality attributes (CQAs). In a conventional approach, four steps are carried out to find such a design space [4,5]. The first step is to perform extensive experiments to determine the relationships between the process parameters and the CQAs.

The second step is to assess the impact of the process parameters on the CQAs (through design of experiments analysis) and select the process parameters that have a medium/high impact on the

CQAs. The third step involves the employment of response surface modeling and optimization to establish a design space graphically. The final step is to run confirmatory experiments to verify the design space that will be submitted to the regulatory agency for assessment and approval. A few recent industrial applications of such a traditional method have been reported by Kumar et al. (2014) [6] and Chatzizaharia and Hatziavramidis (2015) [7].

However, establishing the design space with this approach has significant disadvantages. Pharmaceutical processes are expensive and associated raw materials may be costly. Furthermore, extensive experimentation is time consuming. Therefore, there are limits on the number of experiments that can be performed in practice. Recently, data-driven approaches like Bayesian methods [8] and multivariate statistical techniques such as PCA and PLS [3,5,9] have been used to better manage the extent of these costs, however, these techniques require significant, high quality data [10]. Alternatively, we can use mechanistic models that intrinsically contain relationships between process parameters, uncertain variables, and critical quality attributes. This model-based approach allows for more informative and targeted experiments to be performed during design space formulation.

In a model-based approach we consider process parameters θ_p which include both design decisions and fixed process decisions that do not change during operation (e.g., reactor dimensions, feed conditions). Assuming deterministic system behavior, the *deterministic* design space can be easily found by performing simulations over the space of these process parameters and checking the critical quality attributes at each of these points. However, uncertainty in model parameters plays an important role and cannot be ignored. Uncertain model parameters θ_m (e.g., kinetic rate constants, heat transfer coefficients) are typically estimated by maximum-likelihood or Bayesian techniques based on experimental data. In addition to point estimates of the parameters, such approaches provide an estimate of the distribution of those uncertain model parameters (e.g., covariance matrix). The uncertainty arising from this estimation propagates to uncertainty in the acceptability of the CQAs [4]. Accounting for this uncertainty, the *probabilistic* design space captures the region in the process parameter space where product quality is assured within a given probability over the uncertain parameters.

One approach to determine the probabilistic design space is through Monte-Carlo simulation. The space of process parameters is first discretized (e.g., a fine uniform grid), and for each point in the process parameter space an ensemble of simulations is performed using sampled values for the uncertain model parameters. For every sampled simulation, the CQAs can be checked, recording success or failure and, over the entire ensemble, the probability that the CQAs are acceptable can be computed for each particular point in the process parameter space. This approach and similar sample-based approaches have been shown to be effective [11–14], however, they are computationally expensive since simulations are performed for each sample in the Monte-Carlo simulations for every point in the discretized process parameters. There is a need for approaches with improved computational efficiency to address larger uncertainty and process parameter spaces.

The concept of the design space in the pharmaceutical industries is very similar to flexibility analysis [15] from the chemical process industry. They share a similar goal of quantifying the operational flexibility for manufacturers. Halemane and Grossmann (1983) [16], Swaney and Grossmann (1985) [17,18], and Grossmann and Floudas (1987) [19] introduced multi-level optimization formulations to assess flexibility of chemical processes. The flexibility test formulation maximizes the violation of the inequality constraints over a predetermined region in the uncertain parameters. This provides a check of whether or not operation and product quality constraints are satisfied over the entirety of that region [16,19,20]. The flexibility index formulation extends the idea and solves for this region directly. It seeks to find the largest hyperrectangle in the space of the uncertain parameters where the set of inequality constraints is guaranteed to be satisfied [17,18]. Solving these multi-level optimization problems can be very challenging, and early work focused on algorithms for improving efficiency by assuming that the worst-case behavior occurred at vertices of the parameter space [16–18]. An active-set strategy was later proposed that could identify solutions at points that were not necessarily vertices [19,21]. This approach replaced the inner problem (over the control variables) with explicitly first-order optimality conditions and was globally valid only under certain problem assumptions. This limitation was later overcome with an approach that guaranteed global optimality in the general non-convex case (through relaxation) [15].

For a linear system with model parameter uncertainty, a stochastic flexibility index formulation that exploits the probabilistic structure of the problem is presented by Pistikopoulos and Mazzuchi (1990) [22]. Extensions and variations of the flexibility test and flexibility index formulations have been proposed that optimize over both the design and the operations. One example is a two-level formulation that optimizes a certain process performance metric (such as the production rate or the profit) while maximizing the flexibility region for a given design. Many researchers have made significant contributions to the solution to such a problem, including Mohideen et al. (1996) [23], Bahri et al. (1997) [24], Bernardo and Saraiva (1998) [25], and Samsatli et al. (2001) [26].

In this paper, we propose flexibility test and flexibility index formulations within two algorithms to compute the probabilistic design space with improved computational efficiency over traditional Monte-Carlo approaches based on exhaustive simulation. In the first approach, process parameters θ_p are still discretized and, for each fixed point on the process parameter grid, the Monte-Carlo simulations are replaced with a flexibility index formulation. The flexibility index formulation computes a region in the uncertainty space over which the inequalities (e.g., acceptability of the CQAs) are guaranteed to be satisfied. To extend this to the *probabilistic* design space one could employ sample-based approaches or chance constraints (which would increase complexity and computational effort). Instead, we overlay simple statistical testing with the flexibility analysis and solve for the largest region in θ_m that satisfies the CQAs and then determine the probability that a realization of θ_m will lie in this region. We further propose a second approach that pushes these ideas further by solving for the probabilistic design space in θ_p directly. We extend the flexibility test formulation to include a statistical confidence constraint on the uncertain parameters and a hyperrectangle constraint on the process parameters. This approach removes the need to discretize the process parameters and reduces computational time significantly, however, it produces more conservative results since the relative dimensions (but not the size) of the design space is fixed. The results of both of these approaches are validated against the Monte-Carlo sampling approach [4].

The rest of this article is organized as follows. In Section 2, we describe the Monte-Carlo approach from [4], provide background on the flexibility test and flexibility index problems, and then present the proposed approaches for computing the probabilistic design space with extensions to the flexibility analysis concepts. In Section 3, we demonstrate the approach on a small case study as well as the industrial Michael addition reaction case provided by the Eli Lilly and Company [27]. These case studies are used to compare the effectiveness of the new approaches with the Monte-Carlo simulation based approach. Discussions and conclusions are presented in Section 4.

2. Problem Formulation and Solution Approach

In this section, we will first describe the probabilistic design space problem and the Monte-Carlo solution approach (from [4]). We will then briefly introduce the concept of the flexibility test and flexibility index formulations and introduce the two proposed approaches to compute the probabilistic design space more efficiently.

2.1. Probabilistic Design Space

Recall that θ_p are the process parameters; these are the process design variables and processing decisions that are fixed during operation (e.g., fixed temperatures, pressures, or feed conditions), and θ_m are the uncertain parameters in the mechanistic model (e.g., reaction rate constants, heat transfer coefficients). It is assumed that the uncertain model parameters have been estimated (e.g., from experimental data), and that nominal values and the covariance matrix are available. Using this notation, the model for the process is given by,

$$h(\theta_p, x, \theta_m) = 0$$

where x are the internal state variables computed from the model. The critical quality attributes (CQAs) can be represented by the set of inequalities,

$$g(\theta_p, x, \theta_m) \leq 0.$$

With these definitions, the *deterministic* design space is the region in θ_p over which the CQAs are satisfied while using nominal values for the uncertain parameters. The probabilistic design space considers uncertainty in the model parameters. It is characterized as the region in θ_p over which the CQAs are satisfied with a given probability, where this probability is computed over the distribution of the uncertain parameters. Note that the characterization of the probabilistic design space in García Muñoz et al. (2015) [4] does not include adjustable control action to increase the size of the probabilistic design space. For some pharmaceutical processes there is minimal online measurement and control of the CQAs, and the process is instead carried to completion, followed by testing of the final product. Furthermore, while traditional flexibility test and index formulations solve directly for "optimal" control values, the underlying control laws may not be easy to implement in practice. Therefore, consistent with the definition in García Muñoz et al. (2015) [4], we assume that any online control is included directly in the model equations and not available for the optimization.

2.2. *Probabilistic Design Space using Monte-Carlo*

The Monte-Carlo approach for determining the probabilistic design space is shown below in Algorithm 1 [4]. Let Θ_p be the set of discretized points for the process parameters θ_p (usually over a uniform grid). For each of the points in this grid, the uncertain parameters are sampled, and the ensemble of simulations is performed. The CQAs are checked for each of these simulations, and the probability of acceptable operation is computed based on the fraction of samples for which the CQAs are satisfied.

Algorithm 1 Monte-Carlo Probabilistic Design Space Determination

1: Discretize the process parameter space ($\Theta_p = \{\theta_p^i \; \forall \, i\}$)
2: **for** each θ_p^i **do**
3:     **Monte-Carlo Sampling**
4:         Generate samples for uncertain parameters ($\theta_m^j \sim \mathcal{N}(\bar{\theta}_m, C)$)
5:         **for** each θ_m^j **do**
6:             Perform the simulation: solve $h(\theta_p^i, x, \theta_m^j) = 0$ for x^{ij}
7:             Check the CQAs (i.e., are all $g(\theta_p^i, x^{ij}, \theta_m^j) \leq 0$)
8:         Compute probability that CQAs are satisfied for θ_p^i
9: Generate the probability map over all points $\theta_p^i \in \Theta_p$

This approach is effective at determining the probabilistic design space. The grid can be made arbitrarily fine through discretization of the process parameters, and the sampling step has no restriction on the distribution of the uncertain parameters. However, the number of simulations that need to be performed is equal to the number of process parameter discretizations (i.e., grid points in θ_p) times the number of samples used in the Monte-Carlo step. Because of this, the computational cost of the approach can be prohibitive.

The major computational overhead of the Monte-Carlo approach described above is related to the large number of simulations performed due to the discretization of the process parameter space in step 1 and the Monte-Carlo sampling in step 3. In this paper, we propose two approaches that make use of optimization-based flexibility concepts, and in the next section, we provide background information on the flexibility test and flexibility index formulations, followed by a presentation of our approaches for determining the probabilistic design space.

2.3. Flexibility Test and Flexibility Index Background

The flexibility test formulation is an approach to verify that a set of inequality constraints (e.g., feasibility with respect to the CQAs) are satisfied over the entirety of a prespecified range of the uncertain parameters. The formulations are typically written as multi-level programming problems. The flexibility test problem is shown below [16,20].

$$
\begin{aligned}
&\chi(d) = \max_{\vartheta \in T} \min_{z} \max_{k \in K} g_k(d,x,z,\vartheta) \\
&\text{s.t. } h_l(d,x,z,\vartheta) = 0 \qquad l \in L \\
&\quad z^L \le z \le z^U \\
&\quad \vartheta^L \le \vartheta \le \vartheta^U
\end{aligned}
$$

The formulation assumes fixed values for the design variables d. These include traditional design decisions (e.g., reactor dimensions) and any processing decisions that are fixed during operation (e.g., feed concentrations). The equality constraints h_l represent the system model, and the inequality constraints g_k represent the feasibility constraints, capturing product quality requirements or other operational constraints. The variables x represent state variables for the system, and z are control variables. The uncertain model parameters are given by ϑ (e.g., reaction rate constants).

Given a particular fixed design d and specified bounds on the uncertain parameters ϑ, this formulation finds the point in ϑ that maximizes the violation of the inequality constraints. Note that the optimal value may be negative (i.e., there is no violation). Therefore, if the value of $\chi(d) \le 0$, then the design is feasible with respect to the inequalities over the entire uncertainty range. In the traditional treatment, the inner formulation is maximizing over ϑ (i.e., finding the *worst-case* value for the feasibility constraints g_k over the uncertain parameters) while minimizing over the control variables z since they can be adjusted during operation to satisfy (as well as possible) the feasibility constraints.

The flexibility index problem extends this idea and, instead of testing over a given region, directly finds the largest region in the parameter space over which the set of inequality constraints are guaranteed to be satisfied. The flexibility index problem is shown below [17–19,21]:

$$
\begin{aligned}
&F(d) = \max \delta \\
&\quad \text{s.t. } \chi(d) = \max_{\vartheta \in T} \min_{z} \max_{k \in K} g_k(d,x,z,\vartheta) \le 0 \\
&\qquad \text{s.t. } h_l(d,x,z,\vartheta) = 0 \qquad l \in L \\
&\qquad\quad z^L \le z \le z^U \\
&\qquad\quad \vartheta^N - \delta\Delta\vartheta^- \le \vartheta \le \vartheta^N + \delta\Delta\vartheta^+ \\
&\qquad\quad \delta \ge 0
\end{aligned}
$$

Given a feasible nominal parameter value ϑ^N, this formulation seeks to find the largest value of δ where the feasibility constraints are still satisfied. In the formulation above, the flexibility index region is characterized as a hyperrectangle in ϑ with scaled deviations $\Delta\vartheta^+$, $\Delta\vartheta^-$, although other representations of this constraint can be used.

Both formulations shown above are particularly challenging because they contain a multi-level optimization problem, which are difficult to solve directly. Floudas and Grossmann (1987) [21] and Grossmann and Floudas (1987) [19] proposed an active-set strategy based on the idea that $\varphi(d,\vartheta^c) = \min_{z} \max_{k \in K} g_k(d,x,z,\vartheta) = 0$ holds at the solution to the flexibility index problem, and F is given by the smallest δ to the boundaries of the feasible region ($\varphi(d,\vartheta) = 0$). With this approach, the flexibility index formulation is transformed into a mixed-integer minimization problem that selects the set of active constraints g_k. Their reformulation handles the inner minimization over z by incorporating the first-order optimality conditions (KKT conditions) of this inner problem directly as constraints in the formulation. This reformulation for the flexibility index problem is given below,

$$
\begin{aligned}
F(\mathbf{d}) = & \min \delta \\
\text{s.t.}\ & h_l(\mathbf{d},\mathbf{x},\mathbf{z},\theta) = 0 && l \in L \\
& s_k + g_k(\mathbf{d},\mathbf{x},\mathbf{z},\theta) = 0 && k \in K \\
& \sum_k \lambda_k \frac{\partial g_k}{\partial x} + \sum_l \eta_l \frac{\partial h_l}{\partial x} = 0 \\
& \sum_k \lambda_k \frac{\partial g_k}{\partial z} + \sum_l \eta_l \frac{\partial h_l}{\partial z} = 0 \\
& \sum_k \lambda_k = 1 \\
& \sum_k y_k = n_z + 1 \\
& \lambda_k - y_k \le 0 && k \in K \\
& s_k - U(1 - y_k) \le 0 && k \in K \\
& \theta^N - \delta\Delta\theta^- \le \theta \le \theta^N + \delta\Delta\theta^+ \\
& \lambda_k, s_k \ge 0, y_k \in \{0,1\} && k \in K \\
& \delta \ge 0
\end{aligned}
$$

where n_z is the number of control variables, s_k are non-negative slack variables, λ_k and η_l are Lagrange multipliers, and y_k are binary variables indicating which constraints g_k are active.

In this paper, we are applying the flexibility test and the flexibility index problems to compute the probabilistic design space as described in García Muñoz et al. (2015) [4]. As discussed earlier, their treatment of the probabilistic design space does not consider optimization of the control action to increase the size of the design space. Therefore, it is assumed that there are no controls, or that the control behavior is included explicitly in the model equations. As shown in Floudas (1985) [20] and Grossmann et al. (2014) [28], applying the active-set approach to the flexibility test problem for the case where $n_z = 0$ gives a formulation shown with Equations (1)–(7) below.

$$\chi(d) = \max_{u,x,\vartheta,s,y} u \tag{1}$$

$$\text{s.t.}\ h_l(d,x,\vartheta) = 0 \qquad l \in L \tag{2}$$

$$s_k + g_k(d,x,\vartheta) - u = 0 \qquad k \in K \tag{3}$$

$$s_k - U(1 - y_k) \le 0 \qquad k \in K \tag{4}$$

$$\sum_{k \in K} y_k = 1 \tag{5}$$

$$\vartheta^L \le \vartheta \le \vartheta^U \tag{6}$$

$$y_k \in 0,1, s_k \ge 0 \qquad k \in K \tag{7}$$

This results in a mixed-integer nonlinear programming (MINLP) problem. The new variable u is introduced to represent the largest value of the constraints g_k. Equation (5) ensures that only one of the constraints will be selected. The big-M constraint, Equation (4), along with the bound on s_k ensure that $s_k = 0$ for the selected constraint, and that u is equal to the corresponding g_k. Therefore, at the solution, the objective function will return the largest possible value across all the constraints

g_k. Again, if $\chi(d) \leq 0$ at the solution, then the region defined by ϑ^L and ϑ^U is acceptable to the inequality constraints.

This formulation is significantly easier to address since it does not include the inner minimization over z (i.e., does not include the KKT conditions as constraints). Furthermore, the number of inequalities g_k is generally small and, more importantly, only one g_k needs to be selected. Therefore, this problem is solved efficiently by explicit enumeration of the binary variables (y_k) [28]. Even with these simplifications, however, the solution remains challenging since these problems must be solved to global optimality.

Applying the active-set strategy to the flexibility index problem in the special case where $n_z = 0$ produces a similar transformation as shown below in Equations (8)–(15).

$$F(d) = \min_{\delta,x,\theta,s,y} \delta \tag{8}$$

$$\text{s.t.} \quad h_l(d,x,\theta) = 0 \qquad l \in L \tag{9}$$

$$s_k + g_k(d,x,\theta) = 0 \qquad k \in K \tag{10}$$

$$s_k - U(1-y_k) \leq 0 \qquad k \in K \tag{11}$$

$$\sum_k y_k = 1 \tag{12}$$

$$\theta^N - \delta\Delta\theta^- \leq \theta \leq \theta^N + \delta\Delta\theta^+ \tag{13}$$

$$s_k \geq 0, y_k \in \{0,1\} \qquad k \in K \tag{14}$$

$$\delta \geq 0 \tag{15}$$

For a thorough description of the flexibility index formulation and the active-set approach, see [15,19,20]. In the subsections that follow, we will show how Equations (1)–(7) and Equations (8)–(15) can be adapted within two algorithmic frameworks to compute the probabilistic design space.

2.4. Flexibility Index Formulation in θ_m

In this section, we present our first approach for determining the probabilistic design space using a flexibility index formulation. The flexibility index problem is formulated over the uncertain model parameters θ_m, replacing the Monte-Carlo simulations in Algorithm 1. With this approach, a flexibility index problem is solved for each discretized point in the process parameter space. Although this still requires solving an optimization problem for each of these discretized points, significant computational performance improvement is possible.

This flexibility index formulation is a direct application of Equations (1)–(7) where the process parameters θ_p are treated as fixed design variables (i.e., $d \equiv \theta_p$) and the uncertainty is captured by uncertain model parameters θ_m (i.e., $\vartheta \equiv \theta_m$) as shown below Equations (16)–(23).

$$F\left(\theta_p^i\right) = \min_{\delta_m,\theta_m,x,s,y} \delta_m \tag{16}$$

$$\text{s.t.} \quad h_l\left(\theta_p,x,\theta_m\right) = 0 \qquad l \in L \tag{17}$$

$$s_k + g_k\left(\theta_p,x,\theta_m\right) = 0 \qquad k \in K \tag{18}$$

$$s_k - U(1-y_k) \leq 0 \qquad k \in K \tag{19}$$

$$\sum_{k \in K} y_k = 1 \tag{20}$$

$$\bar{\theta}_m - \delta_m\Delta\theta_m^- \leq \theta_m \leq \bar{\theta}_m + \delta_m\Delta\theta_m^+ \tag{21}$$

$$\delta_m \geq 0 \tag{22}$$

$$y_k \in 0,1, s_k \geq 0 \qquad k \in K. \tag{23}$$

The formulation above is solved for each of the discretized process parameter points $\theta_p^i \in \Theta_p$ (i.e., θ_p^i is fixed), and the formulation is solved directly for δ_m to determine the size of the region in θ_m for each of these points. Since only one g_k needs to be selected, as discussed earlier, this problem is solved efficiently by explicit enumeration of the binary variables (y_k) [28]. Therefore, the problem is solved as a sequence of NLP problems corresponding to each selection of g_k.

Equation (21) characterizes the flexibility region as a hyperrectangle constraint over the uncertain parameters θ_m. Such a hyperrectangle is centered at the nominal point with sides proportional to the expected deviations, $\Delta\theta_m^+$ and $\Delta\theta_m^-$. However, such a formulation fails to account for the correlation between those uncertain model parameters. If we consider the case that θ_m follows a multivariate normal distribution with the mean $\bar{\theta}_m$ and the covariance matrix Σ_{θ_m}, we obtain Equation (24) below.

$$\left(\theta_m - \bar{\theta}_m\right)^T \Sigma_{\theta_m}^{-1} \left(\theta_m - \bar{\theta}_m\right) \leq \delta_m \tag{24}$$

This ellipsoidal constraint can be used to replace the hyperrectangle constraints with a joint confidence region for θ_m. Although this constraint introduces nonlinearity, it is a convex constraint in θ_m. Given that the covariance matrix Σ_{θ_m} is positive semidefinite, Equation (24) may be transformed using an LDL transformation [29]. In our experience, this transformation improves the numerical behavior of these models. Generalization of Equation (24) with an LDL transformation is shown below:

$$\Sigma_{\theta_m}^{-1} = LDL^T \tag{25}$$

$$q^T = \left(\theta_m - \bar{\theta}_m\right)^T LD^{1/2} \tag{26}$$

$$q^T q \leq \delta_m \tag{27}$$

Flexibility index formulations should be written with Equation (21) or (24) (but not both).

This formulation provides a flexibility region over which the constraints are always guaranteed to be satisfied. It remains to provide a link back to the *probabilistic* design space. One approach would be to modify the formulation and consider the use of chance constraints for g_k. However, this would significantly increase complexity and computational effort required to solve the problem. Therefore, we instead take the flexibility region obtained by Equations (16)–(23), and overlay a statistical test based on our knowledge of the mean and covariance of the uncertain parameters, and directly compute the probability that any realization of θ_m will lie within the region defined by δ_m (hyperrectangle or ellipsoid). For the elliptical flexibility region, the cumulative density function (CDF) of the chi-square distribution can be used to calculate the probability directly. However, if we use the hyperrectangle constraint, we still need to integrate the probability density function over θ_m with upper and lower boundaries. Note that this is a simple determination of the probability that a realization from a particular multi-variate normal will lie in the given hyperrectangle, and can be efficiently approximated through sampling.

The overall approach is described in Algorithm 2 below.

Algorithm 2 Probabilistic Design Space with Flex. Index in θ_m

1: Discretize the process parameter space ($\Theta_p = \{\theta_p^i \; \forall \, i\}$)
2: For hyperrectangle region, use Equation (21). Choose $\Delta\theta_m^-$ and $\Delta\theta_m^+$
3: For ellipsoidal region, use Equation (24). The relative scale is set by Σ_{θ_m}
4: **for** each θ_p^i **do**
5: Solve Flexibility Index Problem, Equations (16)–(23)
6: Solve for δ_m using $\theta_p = \theta_p^i$ and Equation (21) or Equation (24)
7: Compute prob. that θ_m will lie in the region identified by δ_m
8: Generate the probability map over all points $\theta_p^i \in \Theta_p$

This algorithm is a direct application of the flexibility index formulation to replace the Monte-Carlo simulations. In step 6, Equations (16)–(23) must be solved globally to guarantee a valid flexibility region. As discussed above, enumeration is used and the solution is found by solving a series of nonlinear programming (NLP) problems, each with a single $y_k = 1$. It should be noted that if a local solver is used, the optimization may solve to a local minimum, resulting in a δ_m that is larger than the global minimum. Unfortunately, this means that the region returned could be larger than the true flexibility region unless a global minimum is found. In the case studies below, we will show results with both local and global solvers for this step.

Once the optimal value for δ_m is obtained, the probability in step 7 is computed directly or through sampling depending on which region is used (i.e., Equation (21) or (24)). Note also that we expect this approach to produce a more conservative representation of the probabilistic design space since it restricts the relative shape of the region in θ_m when solving the flexibility index problem. While there are no points inside the hyperrectangle or ellipsoid that are infeasible, the actual feasible region need not follow this specific shape, and there could be points outside the region that remain feasible with respect to the inequalities. The Monte-Carlo approach would be able to include these points. This will be discussed further in the case studies.

2.5. Flexibility Test Formulation in θ_p

Our second proposed approach is based on iterative solution of an extended flexibility test formulation. The major benefit of this approach is that the flexibility test formulation is written over both θ_p and θ_m, and it solves for the probabilistic design space directly, thereby removing the need to discretize the process parameters altogether. Consider the extended flexibility test formulation shown below with Equations (28)–(36).

$$\chi\left(\delta_p^r\right) = \max_{u,\theta_p,x,\theta_m,s,y} \quad u \tag{28}$$

$$\text{s.t.} \quad h_l\left(\theta_p, x, \theta_m\right) = 0 \qquad l \in L \tag{29}$$

$$s_k + g_k\left(\theta_p, x, \theta_m\right) - u = 0 \qquad k \in K \tag{30}$$

$$s_k - U\left(1 - y_k\right) \leq 0 \qquad k \in K \tag{31}$$

$$\sum_{k \in K} y_k = 1 \tag{32}$$

$$\bar{\theta}_m - \delta_m \Delta\theta_m^- \leq \theta_m \leq \bar{\theta}_m + \delta_m \Delta\theta_m^+ \tag{33}$$

$$\left(\theta_m - \bar{\theta}_m\right)^T \Sigma_{\theta_m}^{-1} \left(\theta_m - \theta_m^N\right) \leq \delta_m \tag{34}$$

$$\bar{\theta}_p - \delta_p^r \Delta\theta_p^- \leq \theta_p \leq \bar{\theta}_p + \delta_p^r \Delta\theta_p^+ \tag{35}$$

$$y_k \in 0,1, s_k \geq 0 \qquad k \in K \tag{36}$$

As with the previous formulation, this problem can be written with Equation (33) or (34) for θ_m (but not both). When solving the formulation, both $\delta_p = \delta_p^r$ and δ_m are fixed. While δ_p is to be determined, δ_m is pre-calculated to ensure that θ_m remains in the a region with a cumulative probability that agrees with the desired confidence level p_c ($p_c = 0.85$ in this paper). The use of the flexibility test formulation simplifies the optimization since there is no multi-level problem to solve. However, we still want to know the largest value of δ_p over which the constraints g are satisfied.Our goal in this approach is to select a value for δ_p that provides the largest region possible. That is the maximum constraint violation is close to zero while still being negative. Here, we choose a simple bisection approach, although other iteration strategies could be considered. With this approach, the discretization of the process parameter space is replaced with a sequence of flexibility test problems to solve for δ_p^*.

This approach is described in Algorithm 3 below.

Algorithm 3 Probabilistic Design Space with Flex. Test in θ_p

1: If choosing hyperrectangle Equation (33) on θ_m
2: set $\Delta\theta_m^-$ and $\Delta\theta_m^+$ for desired relative dimensions
3: If choosing ellipsoidal Equation (34) the relative scale is set by Σ_{θ_m}
4: Compute δ_m based on desired confidence and selection of shape with Equation (33) or (34)
5: Select tolerance ϵ_{tol}
6: Choose $\bar{\theta}_p$, $\Delta\theta_p^-$, and $\Delta\theta_p^+$
7: Let iteration counter $r = 1$
8: Initialize δ_p^L and δ_p^U for the bisection
9: Let $\delta_p^r = \left(\delta_p^L + \delta_p^U\right)/2$
10: Solve Extended Flexibility Test Problem, Equations (28)–(36) for $\chi\left(\delta_p^i\right)$
11: **if** $\chi\left(\delta_p^r\right) > 0$ **then**
12: Let $\delta_p^U = \delta_p^r$
13: **else**
14: **if** $\left|\chi\left(\delta_p^r\right)\right| \leq \epsilon_{tol}$ **then**
15: done: solution is $\delta_p^* = \delta_p^r$
16: **else**
17: Let $\delta_p^L = \delta_p^r$
18: Let $r = r + 1$ and return to 9

In step 5, the desired tolerance must be selected. The algorithm is written to only converge when the $\chi(\delta_p^i) < 0$, so this tolerance is a measure of the distance "inside" the feasibility constraints. For our case studies, this tolerance was set to 1×10^{-5}. In step 6 one must select the nominal point and the relative dimensions of the hyperrectangle for the probabilistic design space in θ_p. This will have a major impact on the size and shape of the final design space. If the physics of the problem are reasonably well understood, then it is often possible to select a nominal point well within the known acceptable region and scale the relative dimensions effectively. In pharmaceutical manufacturing, the relative dimensions are based on routine process parameter variability in equipment [30]. Otherwise, some exploration of the space will need to be performed. In step 8, δ_p^L and δ_p^U must be selected so that $\chi(\delta_p^L) < 0$ and $\chi(\delta_p^U) > 0$ (i.e., the solution is bracketed).

While this approach can be significantly more computationally efficient since the probabilistic design space is computed directly, there are a couple of drawbacks. As discussed in the previous approach, Algorithm 2, we expect the probabilistic design space to be more conservative since θ_m is constrained by a hyperrectangle or ellipsoid. We expect it to be even more conservative since we are also restricting the shape of the probabilistic design space in θ_p to be that of a hyperrectangle. The convenience of a hyperrectangle is useful in providing simple bounds on the process parameters that can be used in the manufacturing batch record to indicate the region around a nominal point that is safely in the design space. When a process parameter falls outside this hyperrectangle, then the full probabilistic design space can be used to determine if the CQAs are still met. Furthermore, Algorithm 2 does not provide a full probability map, but rather a region that is acceptable over a single given value of probability or confidence. In the case studies that follow, Algorithms 2 and 3 will be compared with results from and computational effort required by the Monte-Carlo approach in Algorithm 1.

3. Case Studies

In this section, we compare the results and computational performance of the proposed algorithms for determining the probabilistic design space on two examples. All problems were modeled using Pyomo [31,32], a Python-based optimization modeling environment, and solved using either IPOPT (version 3.11.7) [33] as a local solver or "BARON" (version 16.12.7) [34–36] as a global solver. All timing results were obtained on a 24 core (Intex Xeon E5-2697—2.7 GHz) server with 256 GB of RAM running

Red Hat Enterprise release 6.10. The case studies include one small example for illustrative purposes and an industrial example based on the Michael Addition reaction. For each of these case studies, we compute the probabilistic design space using all three approaches: Algorithm 1, Algorithm 2 and Algorithm 3.

The approach described in Algorithm 1 is used to provide a basis for comparing the computed probabilistic design space and the computational performance. The process parameters are first discretized as described later for each of the individual case studies. Then, for each discretized point, 1000 samples over θ_m are taken according to a known variance-covariance matrix. As described in the algorithm, for each of these samples, the model is simulated, and the fraction of samples that have acceptable values for the CQAs are recorded for each discretized point. The results are then interpolated to create a map of the probabilistic design space.

The approach described by Algorithm 2 replaces the Monte-Carlo sampling but still requires discretization of the process parameters. For all case studies, the process parameters are discretized using the same points as in Algorithm 1 to enable effective comparison. For each of these discretized points, the flexibility index problem is solved as described in the algorithm. Results are shown using both the hyperrectangle connstraint, Equation (21), and the ellipsoidal constraint, Equation (24). For the case of the hyperrectangle, the $\Delta\theta_m^-$ and $\Delta\theta_m^+$ values are chosen to be the standard deviations of the corresponding uncertain parameters θ_m. For each discretized point, once the optimal δ_m is found and the size of the flexibility region in θ_m is identified, we compute the corresponding probability that a realization of θ_m will lie in this region. With these numbers for each discretized point, the probability map in θ_p can be generated and compared with that of the Monte-Carlo approach. Results are shown using both the local solver IPOPT and the global solver BARON. However, recall that the use of a local solver on these formulations, although faster, provides no guarantees, and it is possible that the probabilistic design space could be overestimated.

The approach described in Algorithm 3 is used to solve for the probabilistic design space in θ_p directly. Again, results are included for this approach using both the hyperrectangle Equation (33) and the ellipsoidal Equation (34) for θ_m. For these studies, the value of δ_m value was fixed to correspond to a confidence level of 0.85. For Equation (33), this value was determined iteratively, and for Equation (34), the inverse chi-square distribution was used. The values for $\Delta\theta_p^-$ and $\Delta\theta_p^+$ are chosen to approximately scale δ_p between 0 and 1, and a convergence tolerance of $\epsilon_{tol} = 1 \times 10^{-5}$ was used. As with the previous approach, results are shown using both the local solver IPOPT and the global solver BARON.

3.1. Case Study 1: Simple Reaction

We first consider a simple reaction case provided by Chen et al. (2016) [27]. The reaction kinetics may be described as such:

$$A + B \xrightarrow{k_1} C \tag{37}$$

$$C \xrightarrow{k_2} D + E \tag{38}$$

where A is 3-chlorophenyl-hydrazonopropane dinitrile, B is 2-mercaptoethanol, and the intermediate product C is formed during reaction. The reaction product is D, 3-chlorophenyl-hydrazonocyanoacetamide, with byproduct E, ethylene sulfide. The reaction rates r_i may be calculated by the following equations:

$$r_1 = k_1 c_A c_B \tag{39}$$

$$r_2 = k_2 c_c \tag{40}$$

The uncertain parameters θ_m for this study are the two rates constants (i.e., $\theta_m = \{k_1, k_2\}$). The estimated value for the rate constants is $\hat{k} = [0.31051, 0.026650]$ and the related variance-covariance matrix given by:

$$Cov_{k_i} = \begin{bmatrix} 1.4409 \times 10^{-4} & 3.27 \times 10^{-6} \\ 3.27 \times 10^{-6} & 8.45 \times 10^{-6} \end{bmatrix} \tag{41}$$

In this case, all the reaction rates r_i and component molar concentrations c_i are state variables. The mass balance of a steady state CSTR is given by:

$$0 = F_i^0 - F_i + V = \left(c_i^0 - c_i\right) + \tau \sum_j v_{ij} r_j \tag{42}$$

where F_i^0 and F_i are the inlet and outlet molar flow rates, v_{ij} are the stoichiometric coefficients, r_j are reaction rates, and τ is the residence time. Using the reactions in Equations (37) and (38), we may write the following equations.

$$c_A^0 - c_A + \tau(-r_1) = 0 \tag{43}$$
$$c_B^0 - c_B + \tau(-r_1) = 0 \tag{44}$$
$$c_C^0 - c_C + \tau(r_1 - r_2) = 0 \tag{45}$$
$$c_D^0 - c_D + \tau(r_2) = 0 \tag{46}$$
$$c_E^0 - c_E + \tau(r_2) = 0 \tag{47}$$

In this study, the initial concentrations $\{c_A^0, c_B^0, c_C^0, c_D^0, c_E^0\}$ are set to be $\left\{0.53, 0.53R_{B|A}, 0, 0, 0\right\}$ mol/L. The probabilistic design space is computed over $R_{B|A}$ (the ratio of the concentration of B to A in the feed) and the residence time τ. That is $\theta_p = \{R_{B|A}, \tau\}$. The process parameter space is discretized with $R_{B|A}$ ranging from 4 to 6, and the residence time τ ranging from 350 to 550 s, with 11 and 21 points respectively.

The feasibility of process operation is determined by the CQAs represented with the following inequality constraints:

$$\frac{c_D}{c_A^0 - c_A} \geq 0.9 \Rightarrow 0.9c_A^0 - 0.9c_A - c_D \leq 0 \tag{48}$$
$$\frac{c_D}{c_A + c_B + c_C} \geq 0.2 \Rightarrow c_A + c_B + c_C - 5.0c_D \leq 0 \tag{49}$$

The first equation states that the yield of product D must be greater than 90%. The second equation states that the ratio of the concentration of D to the concentration of unreacted species must be greater than 0.2.

For Equations (28)–(36), the model equations h are represented with Equations (39)–(40) and (43)–(47), and the CQAs are represented with Equations (48) and (49).

All timing results for Case Study 1 are shown in Table 1. Generating the probabilistic design space takes over 45 min using the Monte-Carlo approach and requires significantly more computational effort because of the large number of required simulations. The approaches in Algorithms 2 and 3 are significantly faster taking a little over 35 s and approximately 3 s respectively (using the global solver BARON). The approach using Algorithm 3 is significantly faster, however, recall that it restricts the shape of the probabilistic design space in θ_p to a hyperrectangle as will be seen in the figures later. While results with the local solver IPOPT are faster again (by about a factor of 2), recall that the local solver cannot provide guarantees that the size of the probabilistic design space may be overestimated. For this simple test case, we do not see significant differences in computational timing when formulating Equations (16)–(23) and Equations (28)–(36) with either the hyperrectangle or the ellipsoid constraint.

Table 1. Timing results for Case Study 1 (in seconds).

Approach	IPOPT (Local)	BARON (Global)
Algorithm 1	2745.3	–
Algorithm 2 with Equation (21),	14.2	35.4
Algorithm 2 with Equation (24),	13.0	37.8
Algorithm 3 with Equation (33),	0.767	1.26
Algorithm 3 with Equation (34),	0.865	3.63

Figure 1 shows the probabilistic design space generated by Algorithm 1.

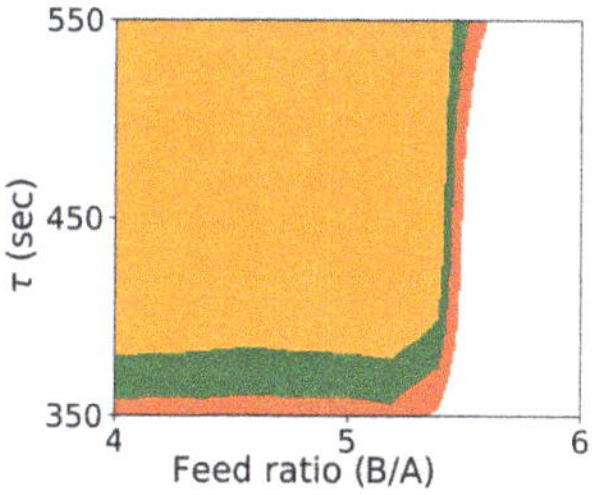

Figure 1. Probabilistic design space for Case Study 1 using Algorithm 1. (orange: $p \geq 0.85$; green: $0.7 \leq p < 0.85$; red: $0.5 \leq p < 0.7$).

The results for the optimization-based flexibility methods are shown in Figure 2, and it includes results for both Algorithm 2 and Algorithm 3, shown for the hyperrectangle and the ellipsoid constraint with both the local solver IPOPT and the global solver BARON.

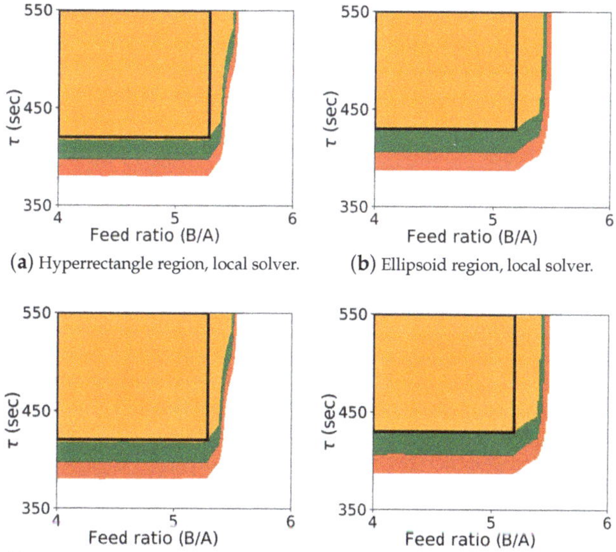

(**a**) Hyperrectangle region, local solver. (**b**) Ellipsoid region, local solver.

(**c**) Hyperrectangle region, global solver. (**d**) Ellipsoid region, global solver.

Figure 2. Probabilistic design space for case study 1 using the flexibility approaches. The colors (orange: $p \geq 0.85$; green: $0.7 \leq p < 0.85$; red: $0.5 \leq p < 0.7$; white: $p < 0.5$) represent the probability map produced by Algorithm 2, and the black rectangle is the probabilistic design space determined by Algorithm 3 with a confidence level of 0.85.

The colored map shows the probabilistic design space obtained using Algorithm 2, and the black rectangle shows the space identified by Algorithm 3 (generated with a single confidence level of 0.85). In this case study, it was known that the upper left corner of the design space corresponded to the "safe" operating region with respect to the CQAs, and the values for $\bar{\theta}_p$, $\Delta\theta_p^-$, and $\Delta\theta_p^+$ could be effectively selected *a priori*. Also, since the shape of the probabilistic design space itself is also rectangular, the differences between the regions from Algorithm 2 and Algorithm 3 are not dramatic. These differences can be more pronounced with other case studies. For this case study, the probabilistic design space identified is similar using both the hyperrectangle and ellipsoid constraints, and the regions identified with the local and global solver are also very similar.

As expected, if we compare these results with the results from Algorithm 1 shown in Figure 1, we see that the design space from the flexibility-based methods is indeed more conservative. Consider results from Algorithm 2. While there are minor differences with respect to $R_{B|A}$, the lower value for τ corresponding to a confidence level of 0.85 is approximately 375 for the Monte-Carlo approach and 425 for the flexibility-based approaches. This is because the shape of the flexibility region in θ_m is restricted. Consider a single point in the process parameter space. Figure 3 shows the results of 1000 simulations (from the Monte-Carlo approach), where the green points are feasible with respect to the CQAs, and red points are not. On this figure, we are also showing the hyperrectangle and ellipsoid generated with Algorithm 2. We can immediately see the impact of restricting the shape. Because of the constraints, the acceptable region for the CQAs in θ_m is not symmetric, and the Monte-Carlo approach is able to identify acceptable points that the flexibility-based approaches are not.

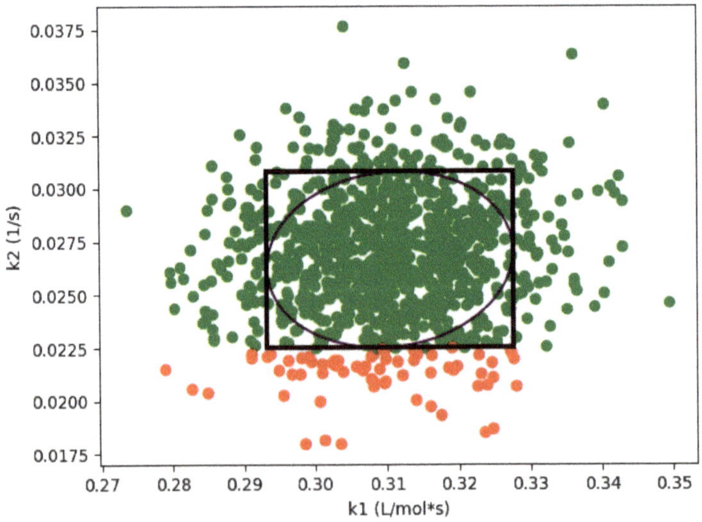

Figure 3. For case study 1, flexibility index produces a more conservative region than Algorithm 1. Green points are feasible, red points are not feasible. This figure demonstrates both ellipsoid and hyperrectangle regions are more conservative at τ = 400.0 and R = 5.0.

3.2. Case Study 2: Michael Addition Reaction

In this section, we consider an industrial case provided by the Eli Lilly and Company—the Michael addition reaction [27] with kinetics described in the following equations:

$$AH + B \xrightarrow{k_1} A^- + BH^+ \tag{50}$$

$$A^- + C \xrightarrow{k_2} AC^- \tag{51}$$

$$AC^- \xrightarrow{k_3} A^- + C \tag{52}$$

$$AC^- + AH \xrightarrow{k_4} A^- + P \tag{53}$$

$$AC^- + BH^+ \xrightarrow{k_5} P + B \tag{54}$$

where AH (Michael donor) and C (Michael acceptor) are starting materials, B is a base, BH^+, A^- and AC^- are reaction intermediates, and P is the product. Reaction rates r_i are defined as follows:

$$r_1 = k_1 c_{AH} c_B \tag{55}$$

$$r_2 = k_2 c_{A^-} c_C \tag{56}$$

$$r_3 = k_3 c_{AC^-} \tag{57}$$

$$r_4 = k_4 c_{AC^-} c_{AH} \tag{58}$$

$$r_5 = k_5 c_{AC^-} c_{BH^+} \tag{59}$$

The rate constants k_i are the uncertain model parameters (i.e., $\theta_m = \{k_1, k_2, k_3, k_4, k_5\}$), and these rate constants have estimated values,

$$\hat{k} = \begin{bmatrix} 49.7796 & 8.9316 & 1.3177 & 0.3109 & 3.8781 \end{bmatrix}$$

and the multivariate normal variance-covariance matrix given by:

$$\begin{bmatrix} 1.005 & -3.428 \times 10^{-4} & -1.006 \times 10^{-3} & -1.523 \times 10^{-3} & 2.718 \times 10^{-3} \\ -3.428 \times 10^{-4} & 0.412 & 7.951 \times 10^{-4} & -3.937 \times 10^{-3} & 2.364 \times 10^{-3} \\ -1.006 \times 10^{-3} & -7.951 \times 10^{-4} & 3.224 \times 10^{-3} & 1.466 \times 10^{-3} & -2.400 \times 10^{-3} \\ -1.523 \times 10^{-3} & -3.937 \times 10^{-3} & 1.466 \times 10^{-3} & 2.746 \times 10^{-3} & -4.102 \times 10^{-3} \\ 2.718 \times 10^{-3} & 2.364 \times 10^{-3} & -2.400 \times 10^{-3} & -4.102 \times 10^{-3} & 7.148 \times 10^{-3} \end{bmatrix} \tag{60}$$

Using a CSTR mass balance over the reactions, Equations (50)–(54), we may write the following equations:

$$c_{AH}^0 - c_{AH} + \tau(-r_1 - r_4) = 0 \tag{61}$$

$$c_B^0 - c_B + \tau(-r_1 + r_5) = 0 \tag{62}$$

$$c_C^0 - c_C + \tau(-r_2 + r_3) = 0 \tag{63}$$

$$c_{A^-}^0 - c_{A^-} + \tau(r_1 - r_2 + r_3 + r_4) = 0 \tag{64}$$

$$c_{AC^-}^0 - c_{AC^-} + \tau(r_2 - r_3 - r_4 - r_5) = 0 \tag{65}$$

$$c_{BH^+}^0 - c_{BH^+} + \tau(r_1 - r_5) = 0 \tag{66}$$

$$c_P^0 - c_P + \tau(r_4 + r_5) = 0 \tag{67}$$

In this study, the initial concentrations $\left\{c_{AH}^0, c_B^0, c_C^0, c_{BH^+}^0, c_{A^-}^0, c_{AC^-}^0, c_P^0\right\}$ are set to be $\{0.3955, 0.3955/R, 0.25, 0, 0, 0, 0\}$ mol/L respectively, where R is the molar ratio between the feed concentration of AH and B. The process parameters include the molar ratio R and the residence time τ (i.e., $\theta_p = \{R, \tau\}$). These process parameters are discretized with R from 10 to 30, and τ from 400 to 1400 min, with 21 and 11 points respectively.

Feasible process operation is determined by the following two CQA constraints:

$$\frac{c_C^0 - c_C - c_{AC^-}}{c_C^0} \geq 0.9 \Rightarrow c_C + c_{AC^-} - 0.1 \times c_C^0 \leq 0 \tag{68}$$

$$c_{AC^-} \leq 0.002 \tag{69}$$

The first constraint states that the conversion of feed C must be greater than 90%, and the second states that the concentration of AC^- in the outlet must be less than 0.002 mol/L.

The model equations h are represented with Equations (55)–(59) and (61)–(67), and the CQAs are represented with Equations (68) and (69).

The timing results for this case study can be found in Table 2.

Table 2. Timing results for Case Study 2 (in seconds).

Approach	IPOPT (Local)	BARON (Global)
Alg. 1	6116.2	–
Alg. 2 with Equation (21),	16.3	203
Alg. 2 with Equation (24),	18.0	–
Alg. 3 with Equation (33),	1.26	21.7
Alg. 3 with Equation (34),	1.65	245.3

Here we see similar results as with the first case study. The flexibility-based methods are significantly faster than the Monte-Carlo approach. As before, Algorithm 3 was about an order of magnitude faster than Algorithm 2. However, here we also see one of the challenges of the global optimization approaches. For the ellipsoidal constraint with Algorithm 2, BARON failed to converge for a small number of points, and therefore, timing results are not reported for this case. When using BARON with the ellipsoidal constraint in Algorithm 3, the gap did not close within the specified time limit on two iterations of the bisection method. However, when the maximum allowed time was reached for these two points, both the upper and lower bounds on the objective value were negative, signifying operational feasibility. The LDL transformation of the ellipsoid constraint was used in both formulations during global optimization.

The probabilistic design space generated from the Monte Carlo procedure and from the flexibility-based approaches is shown in Figure 4. As with case study 1, this figure includes results for both Algorithm 2 and Algorithm 3. Since BARON did not solve with the ellipsoidal constraint using Algorithm 2, this figure also includes the Monte-Carlo results in the subfigure (d).

As before, comparing the computed probabilistic design space, we see that Algorithm 2 is more conservative than the Monte-Carlo approach. Here, however, we see the more significant differences between the flexibility-based methods. The rectangular region produced by Algorithm 3 correctly lies inside the probabilistic design space produced by Algorithm 2. But, since the actual probabilistic design space is not rectangular, the rectangular region produced by Algorithm 3 significantly underestimates the size of the region. In some applications, the region that is to be reported may be defined with simple bounds on process parameters, and the rectangular region produced by Algorithm 3 will be sufficient. For other applications, this underestimation may be too dramatic, and extensions of this approach may need to be used to find a larger region (e.g., shifting the nominal point and producing multiple overlapping rectangles).

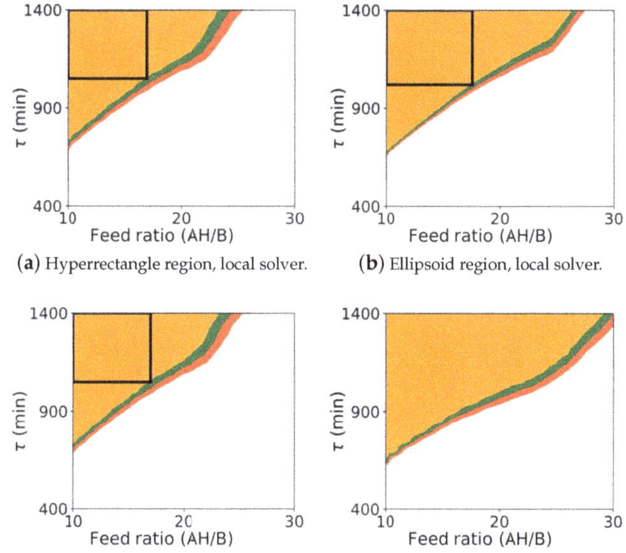

(**a**) Hyperrectangle region, local solver. (**b**) Ellipsoid region, local solver.

(**c**) Hyperrectangle region, global solver. (**d**) Monte-Carlo approach.

Figure 4. Probabilistic design space for case study 2 (Michael Addition) using the flexibility approaches and the Monte-Carlo approach. The colors (orange: $p \geq 0.85$; green: $0.7 \leq p < 0.85$; red: $0.5 \leq p < 0.7$; white: $p < 0.5$) represent the probability map produced by Algorithm 2, and the black rectangle is the probabilistic design space determined by Algorithm 3 with a confidence level of 0.85. The ellipsoid region is not shown with the global solver since the global solver did not converge for those cases.

4. Discussion and Conclusions

A key component of the QbD initiative in the pharmaceutical industry is the identification of the probabilistic design space defined as the region in the space of the process parameters over which the critical quality attributes of the product are acceptable. Traditional Monte-Carlo approaches have been used to compute the probabilistic design space by discretizing the process parameters and performing simulations over hundreds (or more) of samples from the uncertain parameters.

Here, we proposed an optimization-based framework to define the probabilistic design space of a pharmaceutical process with model uncertainty using concepts from flexibility analysis [15,19]. Specifically, we proposed two methods. The first, Algorithm 2, is a direct application of the flexibility index formulation. This approach still discretizes the process parameters θ_p, but replaces the Monte-Carlo simulations with a flexibility index formulation in the uncertain parameters θ_m. The second approach solves for the probabilistic design space in θ_p directly, removing the need to discretize the process parameter space as well. Both these approaches showed significant improvement in computational performance over the Monte-Carlo approach, with Algorithm 3 being another order of magnitude faster than Algorithm 2. While the Monte-Carlo approach can be easily run in parallel, note that Algorithm 2 can also be run in parallel over the discretized points in θ_p. Given the difference in solution time between the Monte-Carlo approach and Algorithm 3, it would take significant HPC resources to make the Monte-Carlo approach faster.

However, the size of the probabilistic design space produced by the flexibility-based approaches is more conservative since they restrict the shape of the confidence region in θ_m and, in the case of Algorithm 3, the shape of the probabilistic design space itself. It will depend on the specific application to determine if this trade-off is acceptable or not. Also, extensions of the flexibility test and flexibility index approaches could be used to reduce this effect.

Due to the problem definition, the formulations presented did not make use of online control action to increase the size of the probabilistic design space. The flexibility test and flexibility index formulations do provide rigorous treatment of controls [15,19,28], and future work will explore this aspect.

With case study 2 (Michael addition reaction), the global solver did not fully converge for either Algorithm 2 or Algorithm 3 when the ellipsoidal constraint in θ_m was used. While this constraint is convex, it was presented to the solver as a sum of bilinear terms, and it is possible that solver tuning or a straightforward outer-approximation would yield improved behavior.

As these problems become larger, performance of the global optimization step will become the primary bottleneck. One approach to improve performance is to instead solve a relaxation of Equations (16)–(23) or Equations (28)–(36). This will produce a design space that is more conservative, but the relaxations could be progressively refined (e.g., piecewise outer approximation) to manage the trade-off between the size of the design space and the computational effort of the approach.

This paper has shown that the concepts of flexibility analysis, and specifically the flexibility test and flexibility index formulations, can be used to compute probabilistic design spaces much more efficiently. Furthermore, there have been many advances in flexibility analysis that could be further applied to improve scalability and reduce conservativeness when estimating the design space.

Author Contributions: Author contributions to this research are as follows: conceptualization, C.L., S.V. and S.G.M.; methodology, D.L., S.X., S.V., J.S.R., S.G.M. and C.L.; results and software, D.L., J.S.R., S.X., and C.L.; validation, D.L., S.X., J.S.R., S.V. and C.L.; writing-original draft preperation, S.X., D.L., and C.L.; writing-review and editing, D.L., S.X., C.L., J.S.R., S.V., and S.G.M.; visualization, D.L., S.X. and C.L.; supervision, S.V., C.L. and S.G.M.; project administration, S.V. and C.L.; funding acquisition, C.L., S.V. and S.G.M.

Funding: This research was funded by Eli Lilly and Company through the Lilly Research Award Program (LRAP).

Acknowledgments: This work is supported by Eli Lilly and Company through the Lilly Research Award Program (LRAP). In addition, we would like to acknowledge the help of my colleagues Michael Bynum and Jianfeng Liu. Sandia National Laboratories is a multimission laboratory managed and operated by National Technology & Engineering Solutions of Sandia, LLC, a wholly owned subsidiary of Honeywell International Inc., for the U.S. Department of Energy's National Nuclear Security Administration under contract DE-NA0003525. *Disclaimer:* This paper describes objective technical results and analysis. Any subjective views or opinions that might be expressed in the paper do not necessarily represent the views of the U.S. Department of Energy or the United States Government.

Conflicts of Interest: The authors declare no conflicts of interest.

References

1. Food and Drug Administration. *Pharmaceutical cGMPs for the 21st Century—A Risk-Based Approach*; Technical Report; U.S. Department of Health and Human Services, Food and Drug Administration, Center for Drug Evaluation and Research (CDER): Rockville, MD, USA 2004.
2. Food and Drug Administration. *Guidance for Industry Q8 Pharmaceutical Development*; Technical Report August; U.S. Department of Health and Human Services, Food and Drug Administration, Center for Drug Evaluation and Research (CDER): Rockville, MD, USA, 2009.
3. Facco, P.; Dal Pastro, F.; Meneghetti, N.; Bezzo, F.; Barolo, M. Bracketing the design space within the knowledge space in pharmaceutical product development. *Ind. Eng. Chem. Res.* **2015**, *54*, 5128–5138. [CrossRef]
4. García Muñoz, S.; Luciani, C.V.; Vaidyaraman, S.; Seibert, K.D. Definition of design spaces using mechanistic models and geometric projections of probability maps. *Org. Process Res. Dev.* **2015**, *19*, 1012–1023. [CrossRef]
5. Huang, J.; Kaul, G.; Cai, C.; Chatlapalli, R.; Hernandez-Abad, P.; Ghosh, K.; Nagi, A. Quality by design case study: An integrated multivariate approach to drug product and process development. *Int. J. Pharm.* **2009**, *382*, 23–32. [CrossRef] [PubMed]
6. Kumar, S.; Gokhale, R.; Burgess, D.J. Quality by Design approach to spray drying processing of crystalline nanosuspensions. *Int. J. Pharm.* **2014**, *464*, 234–242. [CrossRef] [PubMed]
7. Chatzizaharia, K.A.; Hatziavramidis, D.T. Dissolution efficiency and design space for an oral pharmaceutical product in tablet form. *Ind. Eng. Chem. Res.* **2015**, *54*, 6305–6310. [CrossRef]

8. Peterson, J.J. A Bayesian approach to the ICH Q8 definition of design space. *J. Biopharm. Stat.* **2008**, *18*, 959–975. [CrossRef] [PubMed]
9. Thirunahari, S.; Chow, P.S.; Tan, R.B.H. Quality by Design (QbD)-based crystallization process development for the polymorphic drug Tolbutamide. *Cryst. Growth Des.* **2011**, *11*, 3027–3038. [CrossRef]
10. Xu, S.; Lu, B.; Baldea, M.; Edgar, T.F.; Wojsznis, W.; Blevins, T.; Nixon, M. Data cleaning in the process industries. *Rev. Chem. Eng.* **2015**, *31*, 453–490. [CrossRef]
11. Figueroa, I.; Vaidyaraman, S.; Viswanath, S. Model-based scale-up and design space determination for a batch reactive distillation with a dean-stark trap. *Org. Process Res. Dev.* **2013**, *17*, 1300–1310. [CrossRef]
12. Pantelides, C. Pharmaceutical Process & Product Development: What can Process Systems Engineering contribute? In *Future Innovation in Process System Engineering (FIPSE)*; Aldemar-Olympian Village, Western Peloponnese, Greece 2012.
13. Pantelides, C.; Pinto, M.; Bermingham, S. Optimization-based design space characterization using first-principles models. Comprehensive Quality by Design in pharmaceutical development and manufacture. In Proceedings of the AIChE Annual Meeting, Salt Lake City, UT, USA, 7–12 November 2010.
14. Pantelides, C.; Shah, N.; Adjiman, C. Design space, models and model uncertainty. Comprehensive Quality by Design in pharmaceutical development and manufacture. In Proceedings of the AIChE Annual Meeting, Nashville, TN, USA, 8–13 November 2009.
15. Floudas, C.A.; Gümüş, Z.H.; Ierapetritou, M.G. Global optimization in design under uncertainty: Feasibility test and flexibility index problems. *Ind. Eng. Chem. Res.* **2001**, *40*, 4267–4282. [CrossRef]
16. Halemane, K.P.; Grossmann I.E. Optimal Process Design under Uncertainty. *AIChE J.* **1983**, *29*, 425–433. [CrossRef]
17. Swaney, R.E.; Grossmann, I.E. An index for operational flexibility in chemical process design. Part I: Formulation and theory. *AIChE J.* **1985**, *31*, 621–630. [CrossRef]
18. Swaney, R.E.; Grossmann, I.E. An index for operational flexibility in chemical process design. Part II: Computational algorithms. *AIChE J.* **1985**, *31*, 631–641. [CrossRef]
19. Grossmann, I.; Floudas, C. Active constraint strategy for flexibility analysis in chemical processes. *Comput. Chem. Eng.* **1987**, *11*, 675–693. [CrossRef]
20. Floudas, C.A. Synthesis and Analysis of Flexible Energy Recovery Networks. Ph.D. Thesis, Carnegie Mellon University, Pittsburgh, PA, USA, 1985.
21. Floudas, C.; Grossmann, I. Synthesis of flexible heat exchanger networks with uncertain flowrates and temperatures. *Comput. Chem. Eng.* **1987**, *11*, 319–336. [CrossRef]
22. Pistikopoulos, E.N.; Mazzuchi, T.A. A novel flexibility analysis approach for processes with stochastic parameters. *Comput. Chem. Eng.* **1990**, *14*, 991–1010. [CrossRef]
23. Mohideen, M.J.; Perkins, J.D.; Pistikopoulos, E.N. Optimal design of dynamic systems under uncertainty. *AIChE J.* **1996**, *42*, 2251–2272. [CrossRef]
24. Bahri, P.A.; Bandoni, J.A.; Romagnoli, J.A. Integrated flexibility and controllability analysis in design of chemical processes. *AIChE J.* **1997**, *43*, 997–1015. [CrossRef]
25. Bernardo, F.P.; Saraiva, P.M. Robust optimization framework for process parameter and tolerance design. *AIChE J.* **1998**, *44*, 2007–2017. [CrossRef]
26. Samsatli, N.J.; Sharif, M.; Shah, N.; Papageorgiou, L.G. Operational envelopes for batch processes. *AIChE J.* **2001**, *47*, 2277–2288. [CrossRef]
27. Chen, W.; Biegler, L.T.; García Muñoz, S. An approach for simultaneous estimation of reaction kinetics and curve resolution from process and spectral data. *J. Chemom.* **2016**, *30*, 506–522. [CrossRef]
28. Grossmann, I.E.; Calfa, B.A.; Garcia-Herreros, P. Evolution of concepts and models for quantifying resiliency and flexibility of chemical processes. *Comput. Chem. Eng.* **2014**, *70*, 22–34. [CrossRef]
29. Chen, Q.; Paulavičius, R.; Adjiman, C.S.; García Muñoz, S. An Optimization Framework to Combine Operable Space Maximization with Design of Experiments. *AIChE J.* **2018**, *64*, 3944–3957. [CrossRef]
30. Seibert, K.D.; Sethuraman, S.; Mitchell, J.D.; Griffiths, K.L.; McGarvey, B. The Use of Routine Process Capability for the Determination of Process Parameter Criticality in Small-molecule API Synthesis. *J. Pharm. Innov.* **2008**, *3*, 105–112. [CrossRef]
31. Hart, W.E.; Watson, J.P.; Woodrulff, D.L. Pyomo: Modeling and solving mathematical programs in Python. *Math. Prog. Comput.* **2011**, *3*, 219. [CrossRef]

32. Hart, W.E.; Laird, C.D.; Watson, J.-P.; Woodruff, D.L.; Hackebeil, G.A.; Nicholson, B.L.; Siirola, J.D. Pyomo—Optimization Modeling in Python. In *Springer Optimization and Its Applications*; Springer-Verlag: New York, NY, USA, 2017; p. 67.
33. Wächter, A.; Biegler, L.T. On the implementation of an interior-point filter line-search algorithm for large-scale nonlinear programming. *Math. Program.* **2006**, *106*, 25–57. [CrossRef]
34. Sahinidis, N.V. BARON: A general purpose global optimization software package. *J. Glob. Optim.* **1996**, *8*, 201–205. [CrossRef]
35. Tawarmalani, M.; Sahinidis, N.V. *Convexification and Global Optimization in Continuous and Mixed-Integer Nonlinear Programming: Theory, Algorithms, Software, and Applications*; Nonconvex Optimization and Its Applications (Book 65); Kluwer Academic: Dordrecht, The Netherlands; Boston, MA, USA, 2002.
36. Tawarmalani, M.; Sahinidis, N.V. A polyhedral branch-and-cut approach to global optimization. *Math. Program.* **2005**, *103*, 225–249. [CrossRef]

processes

MDPI

Article

Robust Process Design in Pharmaceutical Manufacturing under Batch-to-Batch Variation

Xiangzhong Xie [1,2] and René Schenkendorf [1,2,*]

1 Institute of Energy and Process Systems Engineering, Technische Universität Braunschweig, Franz-Liszt-Straße 35, 38106 Braunschweig, Germany
2 Center of Pharmaceutical Engineering (PVZ), Technische Universität Braunschweig, Franz-Liszt-Straße 35a, 38106 Braunschweig, Germany
* Correspondence: r.schenkendorf@tu-braunschweig.de; Tel.: +49-531-391-65601

Received: 7 June 2019 ; Accepted: 30 July 2019; Published: 3 August 2019

Abstract: Model-based concepts have been proven to be beneficial in pharmaceutical manufacturing, thus contributing to low costs and high quality standards. However, model parameters are derived from imperfect, noisy measurement data, which result in uncertain parameter estimates and sub-optimal process design concepts. In the last two decades, various methods have been proposed for dealing with parameter uncertainties in model-based process design. Most concepts for robustification, however, ignore the batch-to-batch variations that are common in pharmaceutical manufacturing processes. In this work, a probability-box robust process design concept is proposed. Batch-to-batch variations were considered to be imprecise parameter uncertainties, and modeled as probability-boxes accordingly. The point estimate method was combined with the back-off approach for efficient uncertainty propagation and robust process design. The novel robustification concept was applied to a freeze-drying process. Optimal shelf temperature and chamber pressure profiles are presented for the robust process design under batch-to-batch variation.

Keywords: robust process design; batch-to-batch variation; parametric probability-box; point estimate method; pharmaceutical manufacturing; freeze-drying

---

1. Introduction

To implement Quality by Design (QbD) concepts, and to ensure optimally designed processes, over the last two decades model-based process design has become an important tool in pharmaceutical manufacturing and process systems engineering [1–5]. For instance, dynamic process models support recent activities of the Food and Drug Administration (FDA) [6] and the International Council for Harmonisation (ICH) Q11 guideline [7] regarding QbD, and the quantification of process variability [8]. Although uncertainties in process models and parameters are considered, and are frequently incorporated in robust process design concepts [9–13], the applied algorithms are commonly based on perfect uncertainty measures, i.e., using specific probability density functions (PDFs). In addition to probability-based concepts for robust process design, scenario-based methods exist [14–16]. Simulation studies seek the worst-case scenario for which the process is optimized, even when the worst-case scenario rarely occurs in reality, and thus, lead to robust but extremely conservative designs with considerable performance losses. Therefore, robust design concepts for pharmaceutical processes, which aim to maximize process performance while satisfying critical process constraints under probabilistic uncertainties, are preferred, to provide the proper trade-off between process performance and robustness [2,17]. Probabilistic uncertainties, in turn, are the result of noisy experimental data and system identification routines that assume a particular experimental setting, while neglecting batch-to-batch variation effects [17]. In the pharmaceutical industry, the batch operation is the standard operating mode when producing active pharmaceutical ingredients (APIs) and drugs [18], i.e., all materials are charged before the start of processing and discharged at the end of processing. Thus, slight experimental deviations or the degradation of the process equipment might result in batch-to-batch variation [18–20]. The source of batch-to-batch variation is difficult to predict, but can be quantified with process analytical technology (PAT) and multivariate statistical analysis [19,20]. In the literature, it is well-known that batch-to-batch variation causes severe problems in pharmaceutical manufacturing, drug quality, clinical studies, and therapeutics [6,17,21,22]. To lower batch-to-batch variation in pharmaceutical, and to improve QbD measures, analyzing the effect of measurement noise and batch-to-batch variation is essential. The adverse effect of batch-to-batch variation in pharmaceutical manufacturing is studied experimentally for various processing steps, e.g., fermentation, crystallization, and (nanomaterial) formulation [17,19,23]. In model-based process design, in turn, recent studies try to analysis and control batch-to-batch variation effects too [24,25]. For instance, in the case of model-based process design, model parameters can be derived for each batch data set separately. When each batch run is fit individually, batch-to-batch variation leads to different sets of model parameter estimates and parameter uncertainties. Please note that the variability in the model parameters is not exclusively the result of measurement noise, but the joint effect of measurement errors and slight differences in the experimental settings and the raw materials of the batches [6,17]. Thus, simulation studies should consider imprecise uncertainties [26–28] as well. These imprecise uncertainties cannot be described via a single PDF, but via a set of PDFs that is known as the ambiguity set [29,30].

With the ambiguity set, we can distinguish between noise (aleatory uncertainty) and batch-to-batch variation (epistemic uncertainty) [8]. The problem of imprecise uncertainties is also closely related to the Dempster-Shafer theory, where uncertainties are expressed as so-called plausibility functions (maximum amount of probability) and belief functions (minimum amount of probability). The same holds for the probability bounds analysis (PBA), which combines probability theory and interval analysis in probability bounds and probability-box (p-box) concepts [31–33]. Based on these ambiguity set realizations, robust process design aims to incorporate imprecise uncertainties in the framework of robust optimization. For instance, recent studies use p-box design concepts for linear optimization problems in process design [30] and algebraic structural reliability analysis [28]. For dynamical systems, however, uncertainty analysis and propagation are challenging, because the computational costs when standard Monte Carlo simulation techniques are used [34].

In the case of robust process design for nonlinear dynamic systems, highly efficient methods for uncertainty propagation are mandatory [35]. In addition to (quasi-) Monte Carlo simulations and improved sampling techniques [36], surrogate models (e.g., neural networks, Gaussian processes, and polynomial chaos expansion) are used to accelerate uncertainty propagation problems in robust process design, but typically suffer the curse of dimensionality [37–40]; that is, the cost increases exponentially with the number of uncertain model parameters. Alternatively, in our previous work, we demonstrated the usefulness of the point estimate method (PEM) [41] for the robust design of pharmaceutical manufacturing processes [35]. The PEM ensures superior efficiency and workable accuracy for many problems in engineering [41,42]. In the particular case of back-off-based robust design methods, process optimization and uncertainty propagation can be considered sequentially [13,43]. Thus, combining back-off-based robust design concepts with the PEM can lead to a dramatic reduction in computational costs, as demonstrated by Emenike et al. [12] for the synthesis of an API intermediate. To the best of the authors' knowledge, a back-off-based robust process design under batch-to-batch variation has not been reported in the literature. Thus, the purpose of this work is two-fold: (1) We integrate imprecise uncertainties caused by measurement noise and batch-to-batch variation with the p-box approach in model-based process design, and (2) we combine the PEM with a back-off-based approach to solve the underlying p-box robust optimization problem efficiently. In Figure 1, the proposed robustification framework is summarized. The effectiveness of the robust process design under batch-to-batch variation is demonstrated for freeze-drying as a highly relevant pharmaceutical process, where the optimal shelf temperature and chamber pressure profiles are derived for optimal process efficiency and process quality attributes.

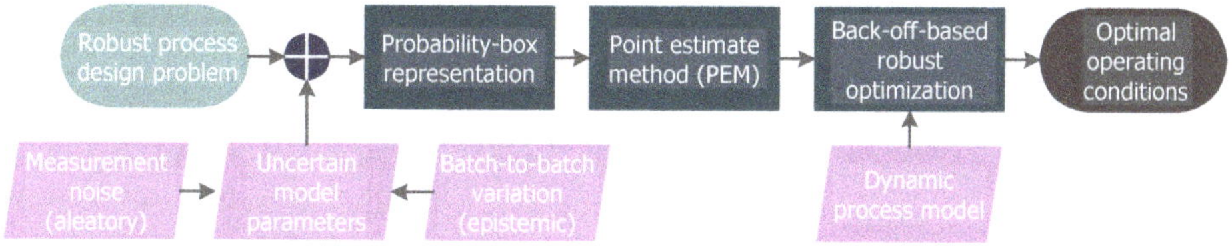

Figure 1. Flowchart of the proposed framework of robust process design under batch-to-batch variation.

The paper is organized as follows. Section 2 covers the basics of the robust process design under batch-to-batch variation. In Section 3, an effective solution strategy with the PEM and the back-off-based design is introduced. Section 4 summarizes the results from the p-box robust process design of a freeze-drying process. Conclusions can be found in Section 5.

2. Robust Process Design

In what follows, the basics of the probability-based process design are summarized. Starting with the standard probability-based robust optimization framework, an extension to imprecise uncertainties representing model parameter uncertainties under batch-to-batch variation is given.

2.1. Probability-Based Robust Optimization

In the literature, various concepts of robust process design exist. Traditional methods for propagating and quantifying model uncertainties are probabilistic and frequently used in robust process design. Here, the interested reader is referred to [9,11,13,16,30,35,37] and references therein. The general structure of the original probability-based robust process design reads as:

$$\min_{\mathbf{x}(\cdot),\mathbf{u}(\cdot)} \quad \Phi(M(\mathbf{x}_{t_f})) \tag{1a}$$

subject to:

$$\dot{\mathbf{x}}_{\mathbf{d}}(t) = \mathbf{g}_{\mathbf{d}}(\mathbf{x}(t), \mathbf{u}(t), \mathbf{p}), \tag{1b}$$

$$\mathbf{0} = \mathbf{g}_{\mathbf{a}}(\mathbf{x}(t), \mathbf{u}(t), \mathbf{p}), \tag{1c}$$

$$\mathbf{x}_{\mathbf{d}}(0) = \mathbf{x}_0, \tag{1d}$$

$$P_v = \mathbf{Pr}[\mathbf{h}_{\mathbf{nq}}(\mathbf{x}(t), \mathbf{u}(t), \mathbf{p}) \geq 0] \leq \varepsilon_{nq}, \tag{1e}$$

$$\mathbf{u}_{min} \leq \mathbf{u} \leq \mathbf{u}_{max}, \tag{1f}$$

where $t \in [0, t_f]$ is the time, $\mathbf{u} \in \mathbb{R}^{n_u}$ is the vector of the control variables, and $\mathbf{p} \in \mathbb{R}^{n_p}$ is the vector of the time-invariant parameters. $\mathbf{x_d} \in \mathbb{R}^{n_{x_d}}$ and $\mathbf{x_a} \in \mathbb{R}^{n_{x_a}}$ are the differential and algebra states; that is, $\mathbf{x} = [\mathbf{x_d}, \mathbf{x_a}] \in \mathbb{R}^{n_x}$. The initial conditions for the differential states are given by $\mathbf{x}_0$. Uncertainties can exist in the parameters and the initial conditions $\boldsymbol{\xi} = [\mathbf{p}; \mathbf{x}_0]$, where the probability space $(\Omega, \mathcal{F}, P)$ is defined with the sample space Ω, the σ-algebra $\mathcal{F}$, and the probability measure P. $\Phi(M(\mathbf{x}_{t_f}))$ denotes the robust formulation of the Mayer objective term $M(\mathbf{x}_{t_f})$ that is used for the nominal process design. Equations (1b) and (1c) are the model equations with $\mathbf{g_d} : \mathbb{R}^{(n_{x_d}+n_{x_a})\times n_u \times n_p} \to \mathbb{R}^{n_{x_d}}$ and $\mathbf{g_a} : \mathbb{R}^{(n_{x_d}+n_{x_a})\times n_u \times n_p} \to \mathbb{R}^{n_{x_a}}$. P_v in Equation (1e) is the probability of violating the inequality constraints $\mathbf{h_{nq}} : \mathbb{R}^{(n_{x_d}+n_{x_a})\times n_u \times n_p} \to \mathbb{R}^{n_{nq}}$. ε_{nq} is the tolerance factor that gives the maximum acceptable probability for constraint violations. $[\mathbf{u}_{min}, \mathbf{u}_{max}]$ are the upper and lower boundaries for the control variables.

For a conventional robust process design, parameters uncertainties $\boldsymbol{\xi}$ are characterized with well-defined probability distributions $F_{\Xi}(\boldsymbol{\xi})$. The probability of constraint violations can be approximated with statistical moments, and thus, the robust inequality constraints in Equation (1e) read as:

$$\mathbf{E}[\mathbf{h_{nq}}] + \beta_{\xi}\mathbf{Var}[\mathbf{h_{nq}}]^{0.5} \leq 0, \tag{2}$$

where $\mathbf{E}[\cdot]$ and $\mathbf{Var}[\cdot]$ indicate the mean and variance calculated over the probability space of $\boldsymbol{\xi}$, and β_{ξ} determines the robust design's conservatism to the variation of the model parameters $\boldsymbol{\xi}$ uncertainties.

2.2. Imprecise Uncertainties

In the case of imprecise uncertainties, the conventional robust process design concept can be generalized with the parametric p-box approach, where the uncertainties of the parameters $\boldsymbol{\xi}$ depend on the hyper-parameters $\boldsymbol{\theta}$ of the parametric probability distributions:

$$F_{\Xi}(\boldsymbol{\xi}) = F_{\Xi}(\boldsymbol{\xi}|\boldsymbol{\theta}), \quad \boldsymbol{\theta} \in \mathcal{D}_{\Theta} \subset \mathbb{R}^{n_\theta}, \tag{3}$$

where $\boldsymbol{\theta}$ is specified by upper and lower bounds, and $\mathcal{D}_{\Theta} = [\theta_1^l, \theta_1^u] \times \ldots [\theta_{n_\theta}^l, \theta_{n_\theta}^u]$ denotes the feasible domain of these hyper-parameters. According to the p-box notation, the probability of a constraint violation can be expressed as a bounded interval $P_v \in [P_v^l, P_v^u]$ with:

$$P_v^l = \min_{\boldsymbol{\theta}} P_v(\boldsymbol{\theta}), \qquad P_v^u = \max_{\boldsymbol{\theta}} P_v(\boldsymbol{\theta}). \tag{4}$$

In the case of the p-box robust process design, the upper probability bound is of interest, to guarantee a safe operation:

$$P_v^u = \max_{\boldsymbol{\theta}}(\mathbf{Pr}[\mathbf{h_{nq}}(\mathbf{x}(t), \mathbf{u}(t), \mathbf{p}) \geq 0|\boldsymbol{\theta}]) \leq \varepsilon_{nq}. \tag{5}$$

If the upper boundary of P_v is lower than or equal to ε_{nq}, then Equation (1e) holds for all realizations of hyper-parameters $\boldsymbol{\theta}$ and $P_v \in [P_v^l, P_v^u]$, respectively. To avoid solving a cumbersome double-loop sampling or optimization problem [31], Equation (5) can be, as for a conventional robust design, also approximated with statistical moments according to:

$$\mathbf{E}_{\boldsymbol{\theta}}[\mathbf{E}_{\boldsymbol{\xi}}[\mathbf{h_{nq}}] + \beta_{\xi}\mathbf{Var}_{\boldsymbol{\xi}}[\mathbf{h_{nq}}]^{0.5}] + \beta_{\theta}\mathbf{Var}_{\boldsymbol{\theta}}[\mathbf{E}_{\boldsymbol{\xi}}[\mathbf{h_{nq}}] + \beta_{\xi}\mathbf{Var}_{\boldsymbol{\xi}}[\mathbf{h_{nq}}]^{0.5}]^{0.5} \leq 0, \tag{6}$$

where β_{θ} determines the conservativeness of the p-box robust design results from the variation of hyper-parameters $\boldsymbol{\theta}$. Note the direct link to PBA, where the first term of Equation (6) refers to the averaged value of the uncertain boundary, and the second term measures the variation. Thus, with $\beta_{\theta} = 2.32$, the 99% confidence interval of the uncertain upper limit is calculated, and the *upper bound of the upper bound* approximates the plausibility function sufficiently. Please note that the 99% confidence interval indicates the interval of the probability distribution, in which the constraints are satisfied, and does not have to be symmetric. Evaluating the plausibility function for a robust process design might result in too-conservative designs under considerable performance loss. Note that the plausibility function assigns the inequality constraints the highest probability. Alternatively, setting $\beta_{\theta} = -2.32$ leads to the lowest probability of the inequality constraint under batch-to-batch variation, and thus, approximates the belief function accordingly. Both strategies, i.e., $\beta_{\theta} = 2.32$ and $\beta_{\theta} = -2.32$, are considered within the PEM-based back-off approach, and the general structure of the double-loop approach is summarized in Figure 2.

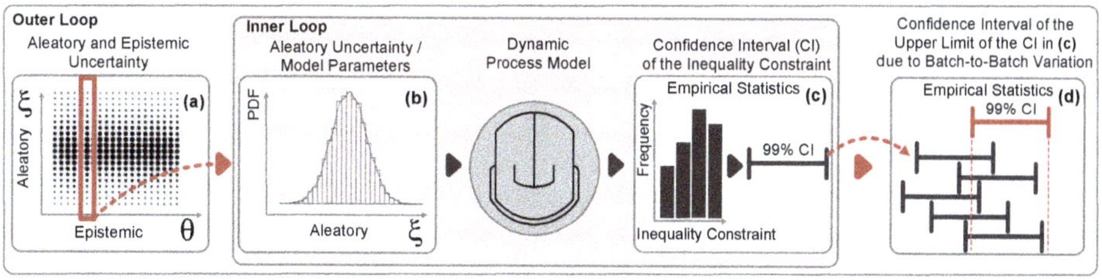

Figure 2. Illustration of the outer and inner loop setting for evaluating the inequality constraint under aleatory and epistemic uncertainty. Sampling from epistemic uncertainty (**a**) results in aleatory uncertainty realization (**b**), which is propagated to the inequality constraint (**c**) via the process model. Re-sampling from the epistemic uncertainty (**a**) helps to quantify the variation in the upper limit of the inequality constraint (**d**).

3. PEM-Based Back-Off Approach

Before introducing the basic notation of back-off-based process design, the concept of the PEM is introduced, and how it can be efficiently used for problems of imprecise uncertainty propagation.

3.1. Point Estimate Method

The conventional robust and p-box robust process design problems are solved with the back-off approach [12], where the back-offs are calculated with Equations (2) and (6), respectively. The statistical moments used in Equations (2) and (6) are approximated with the PEM, as it is more efficient than standard methods for uncertainty propagation [35]. Depending on the underlying parameter distribution, specialized sample points and weight factors w_i can be derived, and evaluated for uncertainty propagation [35,41]. In the case of aleatory parameter uncertainties, $2n_\xi^2+1$ PEM sample points must be used to evaluate Equation (2), assuming Gaussian distributions for the model parameter uncertainties. In detail, the deterministic $2n_\xi^2+1$ sample points are generated by the first three generator functions (**GF**[0], **GF**[$\pm\vartheta$], and **GF**[$\pm\vartheta,\pm\vartheta$]) defined in [41], where ϑ controls the exploration of the n_ξ-dimensional parameter space. Using specific weight factors for each generator function results in the final approximation scheme for the mean value:

$$\mathbf{E}[\mathbf{h_{nq}}] \approx w_0\mathbf{h}^0_{\mathbf{nq}}(\mathbf{GF}[0]) + w_1 \sum_{i=2}^{2n_\xi+1} \mathbf{h}^{\mathbf{i}}_{\mathbf{nq}}(\mathbf{GF}[\pm\vartheta]) + w_2 \sum_{i=2n_\xi+2}^{2n_\xi^2+1} \mathbf{h}^{\mathbf{i}}_{\mathbf{nq}}(\mathbf{GF}[\pm\vartheta,\pm\vartheta]), \tag{7}$$

where $w_0 = 1+\frac{n_\xi^2-7n_\xi}{18}, w_1 = \frac{4-n_\xi}{18}, w_2 = \frac{1}{36}$, and ϑ depends on the specification of the Gaussian distribution [41]. Similarly, the variance can be estimated with the following equation:

$$\begin{aligned}\mathbf{Var}[\mathbf{h_{nq}}] \approx & w_0(\mathbf{h}^0_{\mathbf{nq}}(\mathbf{GF}[0]) - \mathbf{E}[\mathbf{h_{nq}}])^2 + w_1 \sum_{i=2}^{2n_\xi+1} (\mathbf{h}^{\mathbf{i}}_{\mathbf{nq}}(\mathbf{GF}[\pm\vartheta]) - \mathbf{E}[\mathbf{h_{nq}}])^2 \\ & + w_2 \sum_{i=2n_\xi+2}^{2n_\xi^2+1} (\mathbf{h}^{\mathbf{i}}_{\mathbf{nq}}(\mathbf{GF}[\pm\vartheta,\pm\vartheta]) - \mathbf{E}[\mathbf{h_{nq}}])^2.\end{aligned} \tag{8}$$

With Equations (7) and (8), the conventional robust design can be realized, but they ignore the batch-to-batch variation and epistemic uncertainty, respectively. To incorporate epistemic uncertainty, the outer loop of uncertainty propagation must be considered; see Figure 2. To do so, the scaling factor ϑ and weights w_i of the PEM are adapted to uniform probability distributions [35,41]. Thus, for imprecise parameter uncertainties, and the given nested uncertainty propagation problem in Equation (6), $2n_\xi^2+1$ PEM sample points for the model parameters and $2n_\theta^2+1$ PEM sample points for the bounded hyper-parameters $\boldsymbol{\theta}$ are evaluated that result in $4n_\theta^2 n_\xi^2 + 2(n_\theta^2 + n_\xi^2) + 1$ total PEM sample points. Please note that the deterministic sampling scheme from the PEM can be easily parallelized, while ensuring reproducible results.

3.2. Back-Off Realization

For the back-off strategy, the inequality constraint of Equation (1e) is considered in its deterministic form first:

$$\mathbf{h_{nq}}(\mathbf{x}(t), \mathbf{u}(t), \mathbf{p}) \leq 0. \tag{9}$$

To guarantee that the inequality constraint is fulfilled under imprecise uncertainties, a back-off term $\mathbf{b}_c$ to the constraint at the nominal parameter vector $\mathbf{p}$ is introduced as:

$$\mathbf{h_{nq}}(\mathbf{x}(t), \mathbf{u}(t), \mathbf{p}) + \mathbf{b}_c \leq 0. \tag{10}$$

In conventional robust process design with precise parameter uncertainties, the back-off term $\mathbf{b}_c$ can be calculated with the following equation [12]:

$$\mathbf{b}_c = \mathbf{E}[\mathbf{h_{nq}}] + \beta_\xi \mathbf{Var}[\mathbf{h_{nq}}]^{0.5} - \mathbf{h_{nq,nom}}, \tag{11}$$

in which $\mathbf{h_{nq,nom}}$ represents the nominal value of the inequality constraints, and β_ξ determines the robustness that could be obtained with this back-off term. For instance, b_c calculated with $\beta_\xi = 2.32$ could provide a robust design, where 99% of the process realizations under aleatory uncertainty do not violate the inequality constraints.

In comparison with the conventional robust design, the p-box robust process design determines the back-off term b_c with Equation (6) and reads as:

$$\mathbf{b}_c = \mathbf{E}_\theta[\mathbf{E}_\xi[\mathbf{h_{nq}}] + \beta_\xi \mathbf{Var}_\xi[\mathbf{h_{nq}}]^{0.5}] + \beta_\theta \mathbf{Var}_\theta[\mathbf{E}_\xi[\mathbf{h_{nq}}] + \beta_\xi \mathbf{Var}_\xi[\mathbf{h_{nq}}]^{0.5}]^{0.5} - \mathbf{h_{nq,nom}}, \tag{12}$$

in which β_ξ and β_θ decide the robustness of the individual batch and the batch-to-batch variation, respectively. For instance, the back-off term determined with $\beta_\xi = 2.32$ and $\beta_\theta = 2.32$ could provide robust design, with which 99% of different configurations of parameter uncertainties and process realizations will have the desired robust performance; that is, the probability of a constraint violation is smaller than 99%. Different values for β_ξ and β_θ could also be used, depending on the preferred robustness level required for the process under study. For more details regarding the PEM-based back-off design and the selection of proper values for $\mathbf{b}_c$, we refer to our preview works [12,13] and references in those works.

4. Case Study

The freeze-drying process, also known as lyophilization, is used extensively in pharmaceutical manufacturing to stabilize APIs that have limited storage time in aqueous solutions, for example, therapeutic protein formulations and vaccines [10]. The primary drying process, which dominates the overall energy consumption of the lyophilization process, is considered the most critical step [44]. In this study, we analyze the robust process design of the primary drying process in the presence of imprecise parameter uncertainties due to batch-to-batch variations. Thus, we advance our preliminary work on robust process design for the primary drying process which was based on precise parameter uncertainties [45], and we extend recent model-based studies and experimental work on inter-vial heterogeneity in general [46–48].

4.1. Mathematical Model of the Primary Drying Process

The mathematical model of the primary drying process used in this work is adapted from [10,44], and the overall setup is illustrated in Figure 3. The mass transfer equation of vapor, which represents the dynamics of the sublimation process at the sublimation surface, is given as:

$$\frac{\mathrm{d}m_{sub}}{\mathrm{d}t} = A_p \frac{P_i - P_c}{R_p}, \tag{13}$$

where m_{sub} is the mass of ice removed by sublimation. A_p, P_c, and R_p are the cross-sectional area of the product, the chamber pressure, and the dried product resistance to the vapor flux, respectively. The heat used for sublimation is assumed to be equal to the heat transferred from the heating shelf:

$$K_v(T_s - T_B)A_v = \Delta H_s \frac{\mathrm{d}m_{sub}}{\mathrm{d}t}, \tag{14}$$

where K_v, A_v, and T_s are the heat transfer coefficient, the outer cross-sectional area of the vial, and the shelf temperature, respectively.

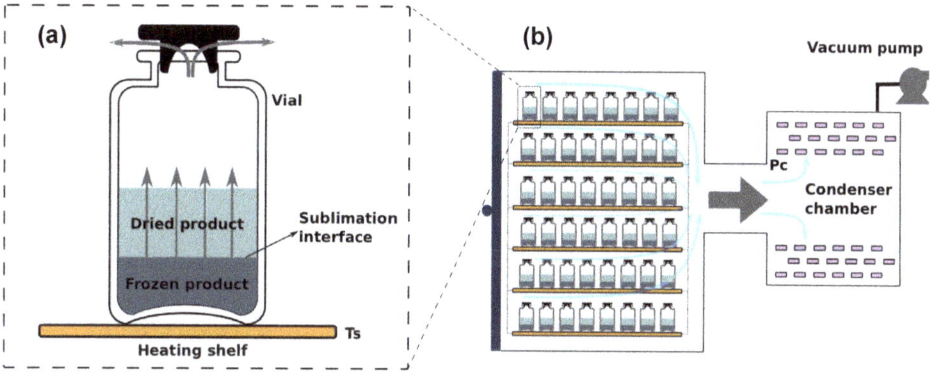

Figure 3. Illustration of the freeze-drying process. Model parameters can represent the dynamic processes of a single vial (**a**) appropriately, but may fail for all vials that are handled in the freeze-dryer (**b**) due to batch-to-batch variations.

P_i is the vapor pressure at the sublimation interface which depends on T_i [49]:

$$P_i = \exp(9.55 - \frac{5720}{T_i} + 3.53\ln(T_i) - 0.00728T_i). \tag{15}$$

T_i is the temperature at the sublimation interface, and is calculated with the energy balance equation given in [44]. K_v, A_v, and T_s are the heat transfer coefficient, the outer cross-sectional area of the vial, and the shelf temperature, respectively. ΔH_s is the heat of sublimation which also depends on T_i [49]. $T_B = T_i + \Delta T$ is the temperature at the bottom of the vial. ΔT is the temperature difference across the frozen layer [10]:

$$\Delta T = \frac{889200\frac{(L_f(P_i - P_c))}{R_p} - 0.0102L_f(T_s - T_i)}{1 - 0.0102L_f}, \tag{16}$$

where L_f is the height of the frozen layer and can be linked to m_{sub} via:

$$m_{sub} = (L_{total} - L_f)\rho_I \epsilon A_p. \tag{17}$$

L_{total}, ρ_I, and ϵ are the total height of the product layer, the density of the ice, and the volume of the ice fraction, respectively. Nominal values and units for the parameters and the initial conditions can be found in [10,44], and the used nominal parameter values are summarized in Table 1.

Table 1. Nominal values of the model parameters and the initial conditions for the primary drying model [10].

Parameters	Symbols	Unit	Nominal Value
cross-sectional area of product	A_p	m^2	3.80×10^{-4}
outer cross-sectional area of the vial	A_v	m^2	4.15×10^{-4}
	$A_{v,n}$	m^2	1.25×10^{-4}
dried product resistance	R_p	m/s	5.57×10^4
heat transfer coefficient	K_v	$J/(m^2sK)$	11.47
	L_{total}	m	0.00658
	ρ_I	kg/m^3	919
	ϵ	–	0.97
	M	kg/mol	0.018
	k	–	1.33
	R	J/(Kmol)	8.314

4.2. Optimal Process Design Strategy

This case study aims to maximize the efficiency of the primary drying step under parameter uncertainties in R_p and K_v, while ensuring the product quality at the same time. The shelf temperature and the chamber pressure are adapted to maximize the total mass of ice removed by sublimation, and to minimize the operating time. To avoid irreversible product damage of the API cake, temperature T_i is maintained to be smaller than the critical collapse temperature T_{crit} = −34 °C [10]. Feasible operation intervals for chamber pressure P_c and shelf temperature T_s are [5, 30] Pa and [−40, 30] °C, respectively.

First, the optimization problem is solved in the absence of parameter uncertainties for the nominal design. Second, precise uncertainties in parameters R_p and K_v are included for the conventional robust process design; that is, batch-to-batch variation effects are ignored. According to [10], we assume that the uncertainties of R_p and K_v follow a Gaussian distribution; i.e., $R_p \sim \mathcal{N}(56{,}000,\ 5600^2)$ and $K_v \sim \mathcal{N}(11.47, 1.15^2)$. Finally, we assume imprecise parameter uncertainties in R_p and K_v for the p-box robust process design as introduced in Section 2. According to the parametric p-box concept, the interval of the hyper-parameters and the type of probability distribution families are listed in Table 2. For the sake of demonstration, the performance of the proposed framework, i.e., precise parameter uncertainties and imprecise parameter uncertainties, are assumed according to the information from [10] and the reference therein.

Table 2. Imprecise uncertainties in model parameters represented as parametric p-boxes.

Parameters	Distribution	Mean Value	Standard Deviation
R_p	Gaussian	[50,000, 80,000]	[5000, 6000]
K_v	Gaussian	[9, 14]	[0.5, 1.5]

The number of model evaluations needed to calculate the back-offs for the p-box robust design is 297 for each iteration, and the back-offs converge at the 4th iteration. The optimization problems were implemented in MATLAB (2017a), and solved within the CasADi framework [50] using the nonlinear programming (NLP) solver IPOPT [51].

4.3. Results and Discussion

In Figure 4, we show the designed profiles of the shelf temperature and the chamber pressure for the nominal, robust, and p-box robust designs of the primary drying step. For the nominal design, T_i is kept at its upper boundary to ensure higher vapor pressure P_i at the sublimation interface, and thus, to accelerate the sublimation process according to Equations (13) and (15). In the beginning, P_c is set to 9.6 Pa to achieve a higher sublimation speed and is decreased gradually to compensate for the influence of the decreasing height of the frozen layer following Equation (16). However, with the existence of (imprecise) parameter uncertainties, the variation of temperature at the sublimation interface will lead to significant violations of the critical temperature which is necessary for maintaining the quality of dried API product. Therefore, results from the robust designs attempt to reduce the temperature of the heating shelf to avoid quality failures of the API cake, while the pressure is maintained at its lower boundary to maximize the efficiency of the freeze-drying process. The shelf temperature for the conventional robust design decreases, while that for the nominal design remains at the upper limit; see Figure 4a. Moreover, the shelf temperature for the p-box robust design (plausibility function, $\beta_\theta = 2.32$) decreases further as the effect of imprecise parameter uncertainties is taken into account with the highest probability. The chamber pressure for the robust and p-box robust designs is also lower than that for the nominal design, and is kept at the minimum to increase the efficiency of the sublimation process; see Figure 4b. The increase in robustness is at the cost of decreased performance; that is, the drying time of the p-box robust design shown in Figure 5 is longer than that for the nominal and conventional robust designs. In addition, the effect of batch-to-batch variation and epistemic uncertainties is also illustrated with the p-box design (belief function, $\beta_\theta = -2.32$). The shelf temperature and chamber pressure profiles are close to those of the nominal case.

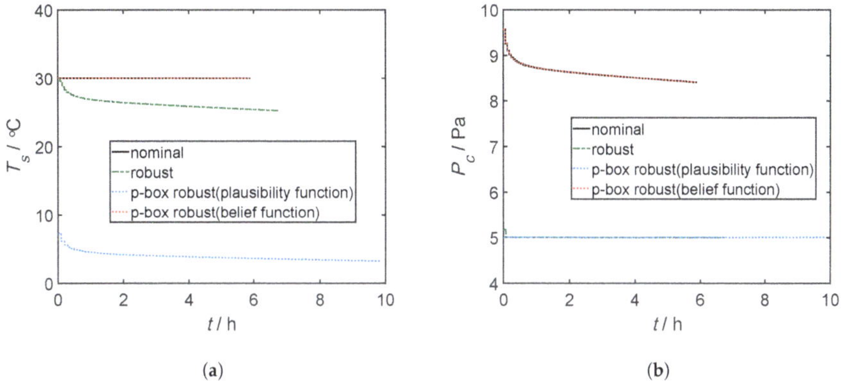

Figure 4. Designed profiles for the shelf temperature T_s (**a**) and the chamber pressure P_c (**b**) of the nominal, robust, p-box (plausibility function) robust, and p-box (belief function) robust designs.

In Figure 5, we illustrate the probability distributions of T_i calculated with the parameter uncertainties of different hyper-parameter realizations for the nominal, robust, and p-box robust designs. Results from the nominal design violate the upper limit of T_i in most scenarios, as indicated in Figure 5a, and thus, the designed shelf temperature and chamber pressure profiles are unlikely to be beneficial for most of the vials that are handled in the freeze-dryer. In Figure 5b, we show that the conventional robust design increases the robustness of the process, but with significant constraint violations when the hyper-parameters are different from the one used for the conventional robust process design, i.e., when there is considerable batch-to-batch variation. In the case of batch-to-batch variation and imprecise parameter uncertainties, the p-box robust design (plausibility function, $\beta_\theta = 2.32$) ensures the lowest number of constraint violations; see Figure 5c. To demonstrate the effect of batch-to-batch variation further, the p-box design (belief function, $\beta_\theta = -2.32$) is given in Figure 5d. Considering the belief function, the p-box design is far from robust, and is close to the nominal design, as indicated in Figure 5d.

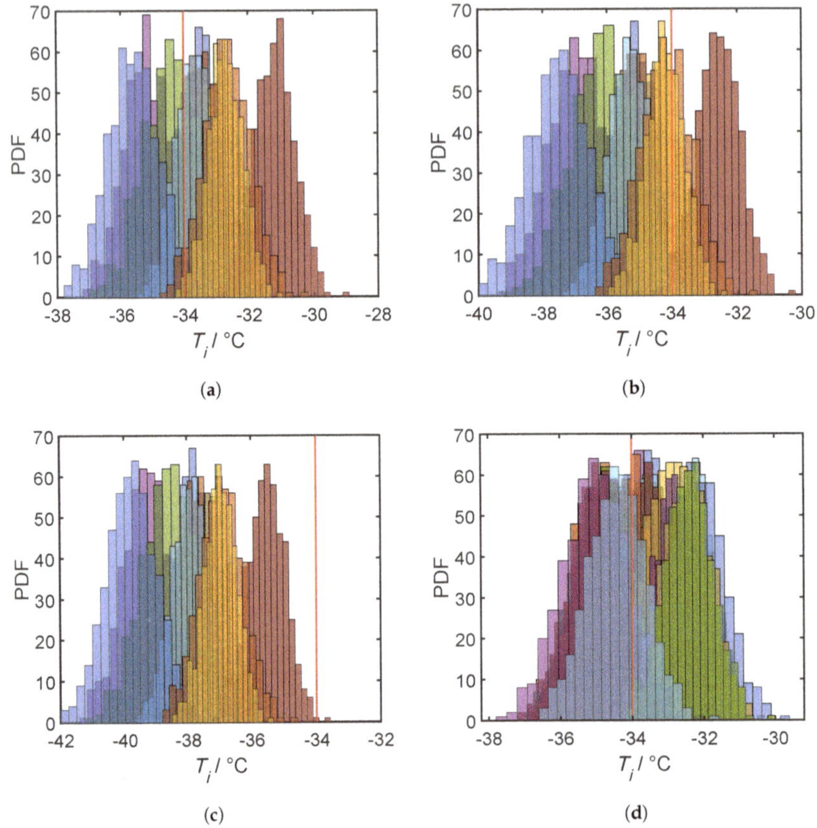

Figure 5. Probability density functions of the temperature at the sublimation interface (T_i) obtained from different probability distributions of parameter uncertainties for (**a**) nominal, (**b**) robust, (**c**) p-box robust (plausibility function), and (**d**) p-box robust (belief function) process designs. The red line is the upper limit of T_i.

The performance of the nominal, robust, and p-box robust designs is further analyzed, and validated for the imprecise parameter uncertainties of R_p and K_v with Monte Carlo simulations. To this end, 1000 realizations of the hyper-parameters (epistemic uncertainty) and 1000 samples of parameters R_p and K_v from each realization (aleatory uncertainty) are generated, which leads to 10^6 samples in total for the double-loop approach (Figure 2). In Figure 6, we summarize the number of constraint violations determined from the 1000 model evaluations with the parameter samples generated from the probability distributions of parameter uncertainties, with a fixed hyper-parameter realization. The normalized histograms of the violation number, in turn, are obtained from the 1000 realizations of the hyper-parameters. Please note that for the sake of validation, we aim for a robust design for the individual batches and batch-to-batch variations. Thus, two parameters, $\beta_{\xi} = 2.32$ and $\beta_{\theta} = 2.32$, are selected, and determine the final robustness level of the designed process. (1) $\beta_{\xi} = 2.32$ (i.e., ε_{nq} is set to 1%) attempts to have a design with which the constraint violation number should be lower than or equal to 10 in the case of 1000 parameter samples in single realization of the hyper-parameters, and (2) $\beta_{\theta} = 2.32$ attempts to guarantee that fewer than or equal to 10 realizations of hyper-parameters will not obey the first desired robustness.

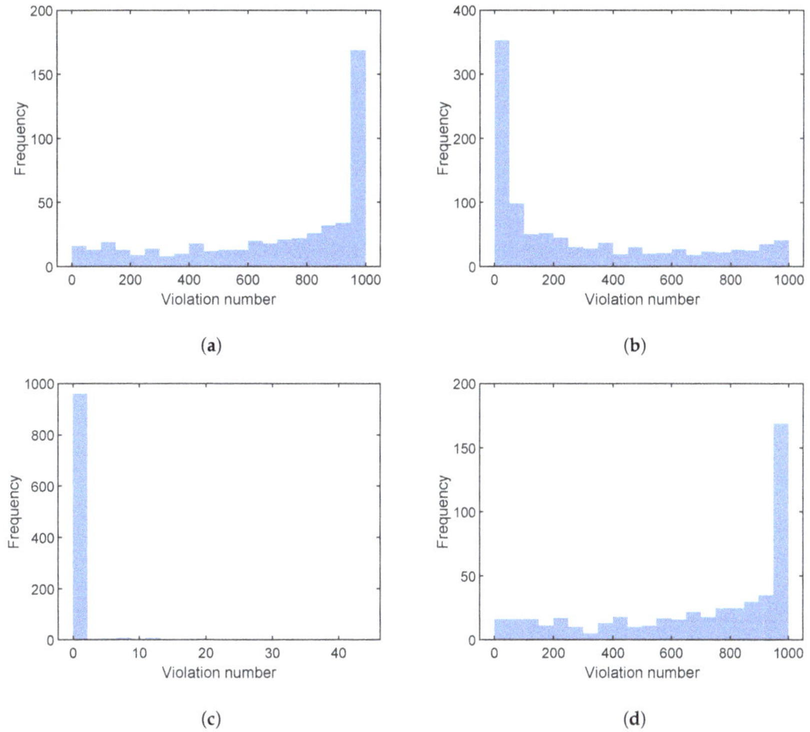

Figure 6. Normalized histograms of constraint violations for the (**a**) nominal, (**b**) robust, (**c**) p-box robust (plausibility function), and (**d**) p-box robust (belief function) process designs.

The probability that the violation number ≥ 10 is the highest for the nominal design; see Figure 6a. As indicated in Figure 6b, the conventional robust process design has a certain effect of, but results in a considerable number of constraint violation, due to the neglected batch-to-batch variation. For the p-box robust design (plausibility function, $\beta_\theta = 2.32$) in Figure 6c, the number of constraint violations fulfills the given specification for almost all sample realizations. Only a small number of samples do not fulfill the given specification of the violation number, which can be attributed to the approximation errors of the PEM used for the inner and outer loops of uncertainty propagation (Figure 2). The effect of the batch-to-batch variation becomes also obvious for the alternative p-box design (belief function, $\beta_\theta = -2.32$) in Figure 6d. The likelihood of a constraint violation increases drastically when compared with the previous p-box design (plausibility function, $\beta_\theta = 2.32$), and has a performance similar to that of the nominal design. To summarize, the probability that the violation number is larger than 10 is equal to 98.6%, 78.9%, 2%, and 96.8% for the nominal, conventional robust, p-box (plausibility function), and p-box (belief functions) designs, respectively. As can be observed, the proposed p-box (plausibility function) approach can handle the imprecise parameter uncertainties and provide a process design which is robust enough for not only an individual batch but also for batch-to-batch variations.

5. Conclusions

In this work, we introduced a p-box robust process design to compensate for batch-to-batch variation and imprecise parameter uncertainties, which were expressed as parametric p-boxes. The notation of the robust inequality constraint was adapted according to the parametric p-boxes, and further approximated with statistical moments that were calculated efficiently. Moreover, combining the point estimate method with a back-off strategy for robust design implementation has been proven to be beneficial in terms of the computational costs for the p-box robust process design concept. The efficiency of the proposed strategy was successfully demonstrated with a freeze-drying process, as a highly relevant pharmaceutical manufacturing process. The results of the p-box robust process design were compared with the results from the nominal and conventional robust designs. The proposed strategy performed quite well, and could ensure the robustness of the inequality constraint, even in the presence of batch-to-batch variation and imprecise parameter uncertainties, respectively. For the p-box design, the two scenarios of plausibility and belief function illustrated the considerable impact of batch-to-batch variation on the optimal process design results. Thus, future work will focus on rigorous sensitivity studies of robust process designs for pharmaceutical processes that have imprecise parameter uncertainties, and systematic analysis of the effect of epistemic and aleatory uncertainties. Moreover, the proposed approach could also be incorporated into advanced control strategies, e.g., model predictive control, to guarantee the robustness of process constraints in the presence of imprecise parameter uncertainties, which is also an interesting aspect for novel Quality by Control concepts.

Author Contributions: X.X. and R.S. designed the study. X.X. performed the and prepared the draft. R.S. provided feedback on the content and participated in writing the manuscript.

Funding: We acknowledge support from the German Research Foundation and the Open Access Publication Funds of the Technische Universität Braunschweig.

Acknowledgments: X. Xie acknowledges the support from the International Max Planck Research School for Advanced Methods in Process and Systems Engineering, MPI Magdeburg.

Conflicts of Interest: The authors declare no conflict of interest.

References

1. FDA. *Pharmaceutical CGMPs for the 21st Century—A Risk-based Approach*; Final Report; U.S. Food and Drug Administration: Rockville, MD, USA, 2014.
2. Gernaey, K.V.; Cervera-Padrell, A.E.; Woodley, J.M. A perspective on PSE in pharmaceutical process development and innovation. *Comput. Chem. Eng.* **2012**, *42*, 15–29. [CrossRef]
3. Wang, Z.; Escotet-Espinoza, M.S.; Ierapetritou, M. Process analysis and optimization of continuous pharmaceutical manufacturing using flowsheet models. *Comput. Chem. Eng.* **2017**, *107*, 77–91. [CrossRef]
4. Metta, N.; Verstraeten, M.; Ghijs, M.; Kumar, A.; Schafer, E.; Singh, R.; De Beer, T.; Nopens, I.; Cappuyns, P.; Van Assche, I.; et al. Model development and prediction of particle size distribution, density and friability of a comilling operation in a continuous pharmaceutical manufacturing process. *Int. J. Pharm.* **2018**, *549*, 271–282. [CrossRef] [PubMed]
5. Emenike, V.N.; Schenkendorf, R.; Krewer, U. A systematic reactor design approach for the synthesis of active pharmaceutical ingredients. *Eur. J. Pharm. Biopharm.* **2018**, *126*, 75–88. [CrossRef] [PubMed]
6. Burmeister Getz, E.; Carroll, K.; Jones, B.; Benet, L. Batch-to-batch pharmacokinetic variability confounds current bioequivalence regulations: A dry powder inhaler randomized clinical trial. *Clin. Pharmacol. Ther.* **2016**, *100*, 223–231. [CrossRef] [PubMed]
7. ICH Expert Working Group. *Development and Manufacture of Drug Substances Q11*; International Council for Harmonisation of Technical Requirements for Pharmaceuticals for Human Use (ICE): Geneva, Switzerland, 2012.
8. Djuris, J.; Djuric, Z. Modeling in the quality by design environment: Regulatory requirements and recommendations for design space and control strategy appointment. *Int. J. Pharm.* **2017**, *30*, 346–356. [CrossRef] [PubMed]
9. Telen, D.; Vallerio, M.; Cabianca, L.; Houska, B.; Van Impe, J.; Logist, F. Approximate robust optimization of nonlinear systems under parametric uncertainty and process noise. *J. Process Control* **2015**, *33*, 140–154. [CrossRef]
10. Mortier, S.T.F.; Van Bockstal, P.J.; Corver, J.; Nopens, I.; Gernaey, K.V.; De Beer, T. Uncertainty analysis as essential step in the establishment of the dynamic Design Space of primary drying during freeze-drying. *Eur. J. Pharm. Biopharm.* **2016**, *103*, 71–83. [CrossRef]

11. Maußer, J.; Freund, H. Optimization Under Uncertainty in Chemical Engineering: Comparative Evaluation of Unscented Transformation Methods and Cubature Rules. *Chem. Eng. Sci.* **2018**, *183*, 329–345. [CrossRef]
12. Emenike, V.N.; Xiangzhong, X.; Schenkendorf, R.; Spiess, A.; Krewer, U. Robust dynamic optimization of enzyme-catalyzed carboligation: A point estimate-based back-off approach. *Comput. Chem. Eng.* **2018**, *121*, 232–247. [CrossRef]
13. Xie, X.; Schenkendorf, R. Stochastic back-off-based robust process design for continuous crystallization of ibuprofen. *Comput. Chem. Eng.* **2019**, *124*, 80–92. [CrossRef]
14. Halemane, K.P.; Grossmann, I.E. Optimal process design under uncertainty. *AIChE J.* **1983**, *29*, 425–433. [CrossRef]
15. Nagy, Z.; Braatz, R. Worst-case and distributional robustness analysis of finite-time control trajectories for nonlinear distributed parameter systems. *IEEE Trans. Control Syst. Technol.* **2003**, *11*, 694–704. [CrossRef]
16. Grossmann, I.E.; Apap, R.M.; Calfa, B.A.; García-Herreros, P.; Zhang, Q. Recent advances in mathematical programming techniques for the optimization of process systems under uncertainty. *Comput. Chem. Eng.* **2016**, *91*, 3–14. [CrossRef]
17. Mockus, L.; Peterson, J.J.; Lainez, J.M.; Reklaitis, G.V. Batch-to-Batch Variation: A Key Component for Modeling Chemical Manufacturing Processes. *Org. Process Res. Dev.* **2014**, *19*, 908–914. [CrossRef]
18. Lee, S.L.; O'Connor, T.F.; Yang, X.; Cruz, C.N.; Chatterjee, S.; Madurawe, R.D.; Moore, C.M.V.; Yu, L.X.; Woodcock, J. Modernizing Pharmaceutical Manufacturing: From Batch to Continuous Production. *J. Pharm. Innov.* **2015**, *10*, 191–199. [CrossRef]
19. Hagsten, A.; Casper Larsen, C.; Møller Sonnergaard, J.; Rantanen, J.; Hovgaard, L. Identifying sources of batch to batch variation in processability. *Powder Technol.* **2008**, *183*, 213–219. [CrossRef]
20. Xiong, H.; Yu, L.X.; Qu, H. Batch-to-batch quality consistency evaluation of botanical drug products using multivariate statistical analysis of the chromatographic fingerprint. *AAPS PharmSciTech* **2013**, *14*, 802–810. [CrossRef]
21. Woodcock, J. The concept of pharmaceutical quality. *Am. Pharm. Rev.* **2004**, *7*, 10–15.
22. Benet, L.Z.; Jayachandran, P.; Carroll, K.J.; Burmeister Getz, E. Batch-to-Batch and Within-Subject Variability: What Do We Know and How Do These Variabilities Affect Clinical Pharmacology and Bioequivalence? *Clin. Pharmacol. Ther.* **2019**, *105*, 326–328. [CrossRef]
23. Mülhopt, S.; Diabaté, S.; Dilger, M.; Adelhelm, C.; Anderlohr, C.; Bergfeldt, T.; Gómez de la Torre, J.; Jiang, Y.; Valsami-Jones, E.; Langevin, D.; et al. Characterization of Nanoparticle Batch-To-Batch Variability. *Nanomaterials* **2018**, *8*, 311. [CrossRef] [PubMed]
24. Nagy, Z.K. Model based robust control approach for batch crystallization product design. *Comput. Chem. Eng.* **2009**, *33*, 1685–1691. [CrossRef]
25. Su, Q.; Chiu, M.S.; Braatz, R.D. Integrated B2B-NMPC control strategy for batch/semibatch crystallization processes. *AIChE J.* **2017**, *63*, 5007–5018. [CrossRef]
26. Baudrit, C.; Dubois, D. Practical representations of incomplete probabilistic knowledge. *Comput. Stat. Data Anal.* **2006**, *51*, 86–108. [CrossRef]
27. Beer, M.; Ferson, S.; Kreinovich, V. Do we have compatible concepts of epistemic uncertainty? In Proceeedings of the 6th Asian-Pacific Symposium on Structural Reliability and its Applications (APSSRA6), Shanghai, China, 28–30 May 2016; pp. 1–10.
28. Schöbi, R.; Sudret, B. Structural reliability analysis for p-boxes using multi-level meta-models. *Probabilistic Eng. Mech.* **2017**, *48*, 27–38. [CrossRef]
29. Ghosh, D.D.; Olewnik, A. Computationally Efficient Imprecise Uncertainty Propagation. *J. Mech. Des.* **2013**, *135*, 051002. [CrossRef]
30. Shang, C.; You, F. Distributionally robust optimization for planning and scheduling under uncertainty. *Comput. Chem. Eng.* **2018**, *110*, 53–68. [CrossRef]
31. Ferson, S.; Troy Tucker, W. Sensitivity analysis using probability bounding. *Reliab. Eng. Syst. Saf.* **2006**, *91*, 1435–1442. [CrossRef]
32. Schöbi, R.; Sudret, B. Uncertainty propagation of p-boxes using sparse polynomial chaos expansions. *J. Comput. Phys.* **2017**, *339*, 307–327. [CrossRef]
33. Bi, S.; Broggi, M.; Wei, P.; Beer, M. The Bhattacharyya distance: Enriching the P-box in stochastic sensitivity analysis. *Mech. Syst. Signal Process.* **2019**, *129*, 265–281. [CrossRef]

34. Smith, R.C. *Uncertainty quantification: Theory, Implementation, and Applications*; siam: Philadelphia, PA, USA, 2013; Volume 12.
35. Xie, X.; Schenkendorf, R.; Krewer, U. Toward a comprehensive and efficient robust optimization framework for (bio)chemical processes. *Processes* **2018**, *6*, 183. [CrossRef]
36. Shi, J.; Biegler, L.T.; Hamdan, I.; Wassick, J. Optimization of grade transitions in polyethylene solution polymerization process under uncertainty. *Comput. Chem. Eng.* **2016**, *95*, 260–279. [CrossRef]
37. Nimmegeers, P.; Telen, D.; Beetens, M.; Logist, F.; Impe, J.V. Parametric uncertainty propagation for robust dynamic optimization of biological networks. *BMC Syst. Biol.* **2016**, 6929–6934.
38. Paulson, J.A.; Mesbah, A. An efficient method for stochastic optimal control with joint chance constraints for nonlinear systems. *Int. J. Robust Nonlinear Control* **2017**, 1–21. [CrossRef]
39. Bhosekar, A.; Ierapetritou, M. Advances in surrogate based modeling, feasibility analysis, and optimization: A review. *Comput. Chem. Eng.* **2018**, *108*, 250–267. [CrossRef]
40. Dias, L.S.; Ierapetritou, M.G. Optimal operation and control of intensified processes—Challenges and opportunities. *Current Opin. Chem. Eng.* **2019**, 8–12. [CrossRef]
41. Lerner, U.N. Hybrid Bayesian Networks for Reasoning about Complex Systems. Ph.D. Thesis, Stanford University, Stanford, CA, USA, 2002.
42. Schenkendorf, R. A general framework for uncertainty propagation based on point estimate methods. In Proceedings of the Second European Conference of the Prognostics and Health Management Society, Nantes, France, 8–10 July 2014.
43. Srinivasan, B.; Bonvin, D.; Visser, E.; Palanki, S. Dynamic optimization of batch processes: II. Role of measurements in handling uncertainty. *Comput. Chem. Eng.* **2003**, *27*, 27–44. [CrossRef]
44. Fissore, D.; Pisano, R.; Barresi, A.A. Advanced approach to build the design space for the primary drying of a pharmaceutical freeze-drying process. *J. Pharm. Sci.* **2011**, *100*, 4922–4933. [CrossRef]
45. Xie, X.; Schenkendorf, R. Robust optimization of a pharmaceutical freeze-drying process under non-Gaussian parameter uncertainties. *Chem. Eng. Sci.* **2019**, *207*, 805–819. [CrossRef]
46. Scutellà, B.; Passot, S.; Bourlés, E.; Fonseca, F.; Tréléa, I.C. How Vial Geometry Variability Influences Heat Transfer and Product Temperature During Freeze-Drying. *J. Pharm. Sci.* **2017**, *106*, 770–778. [CrossRef]
47. Scutellà, B.; Bourlés, E.; Tordjman, C.; Fonseca, F.; Mayeresse, Y.; Trelea, I.C.; Passot, S. Can the desorption kinetics explain the residual moisture content heterogeneity observed in pharmaceuticals freeze-drying process? In Proceedings of the 6th European Drying Conference: EuroDrying 2017, Liège, Belgium, 19–21 June 2017.
48. Scutellà, B.; Trelea, I.C.; Bourlés, E.; Fonseca, F.; Passot, S. Use of a multi-vial mathematical model to design freeze-drying cycles for pharmaceuticals at known risk of failure. In Proceedings of the 21th International Drying Symposium, Valencia, Spain, 18–21 September 2018; Universitat Politècnica València: Valencia, Spain, 2018; pp. 11–14. [CrossRef]
49. Murphy, D.M.; Koop, T. Review of the vapour pressures of ice and supercooled water for atmospheric applications. *Q. J. R. Meteorol. Soc.* **2005**, *131*, 1539–1565. [CrossRef]
50. Andersson, J.; Åkesson, J.; Diehl, M. CasADi: A symbolic package for automatic differentiation and optimal control. In *Recent Advances in Algorithmic Differentiation*; Springer: Berlin/Heidelberg, Germany, 2012; pp. 297–307.
51. Wächter, A.; Biegler, L.T. On the implementation of an interior-point filter line-search algorithm for large-scale nonlinear programming. *Math. Program.* **2006**, *106*, 25–57. [CrossRef]

processes

MDPI

Article

Online Decision-Support Tool "TECHoice" for the Equipment Technology Choice in Sterile Filling Processes of Biopharmaceuticals

Haruku Shirahata, Sara Badr, Yuki Shinno, Shuta Hagimori and Hirokazu Sugiyama *

Department of Chemical System Engineering, The University of Tokyo, Tokyo 113-8656, Japan
* Correspondence: sugiyama@chemsys.t.u-tokyo.ac.jp

Received: 14 June 2019; Accepted: 3 July 2019; Published: 15 July 2019

Abstract: In biopharmaceutical manufacturing, a new single-use technology using disposable equipment is available for reducing the work of change-over operations compared to conventional multi-use technology that use stainless steel equipment. The choice of equipment technologies has been researched and evaluation models have been developed, however, software that can extend model exposure to reach industrial users is yet to be developed. In this work, we develop and demonstrate a prototype of an online decision-support tool for the multi-objective evaluation of equipment technologies in sterile filling of biopharmaceutical manufacturing processes. Multi-objective evaluation models of equipment technologies and equipment technology alternative generation algorithms are implemented in the tool to support users in choosing their preferred technology according to their input of specific production scenarios. The use of the tool for analysis and decision-support was demonstrated using four production scenarios in drug product manufacturing. The online feature of the tool allows users easy access within academic and industrial settings to explore different production scenarios especially at early design phases. The tool allows users to investigate the certainty of the decision by providing a sensitivity analysis function. Further enrichment of the functionalities and enhancement of the user interface could be implemented in future developments.

Keywords: process design; single-use technology; parenteral manufacturing; MATLAB Production Server; software development; multi-objective decision-making

1. Introduction

Biopharmaceuticals represent a growing fraction of pharmaceutical production and can be used for the treatment of many diseases such as cancer, rheumatism, or nephrogenic anemia. Biopharmaceutical production processes consist of drug substance and drug product manufacturing. Drug substance (DS) manufacturing involves the production of the active pharmaceutical ingredient (API) through upstream cell cultivation and purification processes. On the other hand, drug product (DP) manufacturing involves compounding of the API to the final concentration and sterile filling into vials or syringes.

The equipment used in drug manufacturing processes must satisfy certain quality requirements of cleanliness and sterility to be ready for production [1]. New trends of shifting to small-scale and multiple-product production have increased work of change over operations that are conducted to maintain equipment readiness for production between different batches or products. To realize flexible and efficient production, new technologies, e.g., continuous technology or single-use technology (SUT) are applied [2,3], which have increased the number of possible process alternatives. Continuous technology is actively investigated both in small molecule drug manufacturing [4] and biopharmaceutical manufacturing [5] through modeling [6] and experimental approaches [7]. SUT, another newly applied technology, uses disposable resin-made equipment requiring less time for change-over operations and cleaning validation. SUT can replace the conventional multi-use technology (MUT) featuring reusable stainless steel equipment, which requires cleaning and sterilization for change-over.

The choice between process alternatives involving SUT and MUT equipment is multifaceted. The two technologies feature different characteristics in terms of investment, operational risks, and quality challenges. SUT requires lower initial investment but higher running costs to replace the disposable equipment, whereas MUT requires higher initial investment to install the equipment but lower running costs. Maintaining a constant inventory of the sterile manufacturing equipment is an important scheduling decision, which is required to avoid the supply risks of SUT equipment and production interruptions. Various environmental concerns emerge depending on the chosen equipment technology, e.g., emissions form production and disposal of the resin-made equipment versus the utility consumption associated with the cleaning and sterilization processes required for the stainless-steel equipment. In addition, different quality issues arise with each technology, such as chemical compounds leaching from resin material into the drug solution for SUT, and cross-contamination due to failures in the cleaning procedure for MUT. Therefore, the optimal implementation of available technologies requires multi-criteria decision-making. Previous studies have evaluated these technology options using a single evaluation indicator, with a focus on DS manufacturing, such as with economic or environmental evaluations [8–11]. The authors have also presented a framework for the multi-objective evaluation of equipment choice in sterile filling applications of DP manufacturing [12,13]. Another layer of complexity in the decision-making process is the consideration of hybrid equipment technology alternatives combining both resin-made and stainless-steel-made equipment in the same process. One hybrid technology option was considered in a previous work that applies stainless-steel-made fermentation tank and other resin-made equipment [14]. Ha.S., S.B., and Hi.S., part of the authors of this paper, have also previously developed algorithms for the systematic generation of alternatives and technology choice between SUT, MUT, and hundreds of hybrid alternatives [15]. However, to navigate the complex decision-making process, tools are needed to facilitate the generation of various alternatives, the multi-layered comprehensive assessment of the generated alternatives, and the analysis and visualization of the results.

The evolution of decision-support and process design tools has revolutionized the bulk chemical industry. The use of such tools, e.g., Aspen Plus and HYSYS [16,17], for process static and dynamic simulation has allowed the investigation and analysis of complex processes at different design stages. Specific features of the pharmaceutical industry have limited expanding the use of the same tools. Pharmaceutical production is often carried out in relatively small-scale batches and involves more complex chemical and biological interactions where data can sometimes be unavailable. In addition, another difference in pharmaceutical production is the change-over operations required to ensure equipment readiness. In recent years, the maturing understanding of the processes in the biopharmaceutical sector has led to the development of more appropriate design tools for the pharmaceutical industry. The Aspen Batch Process Developer is a recipe-driven process simulator used for the modeling and design of batch processes that enable economic and environmental evaluation [18]. SIMBIOPHARMA is a prototype tool developed for the assessment of equipment technology options and production strategies with focus on DS manufacturing [19]. Other commercial tools are also available such as BioSolve [20] and Hakobio [21]. BioSolve is a stand-alone cost evaluation tool, while Hakobio is an online tool for plant layout design and estimated area calculation especially with disposable equipment with a limited analysis function inside the process. On the other hand, DP manufacturing has not been fully addressed by such tools, due to the different nature of the processes involved. DP processes combine physical and chemical processes at compounding and filtration with other mechanical assembly line processes, like processes at the sterile filling stage. A decision support and process design tool is still required in this field.

Our current work presents a decision-support tool for DP manufacturing processes of biopharmaceuticals, considering the choice of SUT, MUT, and hybrid alternatives as equipment materials to be the key decision. The tool provides a comprehensive multi-objective evaluation of several critical aspects including economic, environmental, quality, and supply robustness considerations. The tool "**TECH**oice" (/tɛktʃɔɪs/), is derived from the combination of "technology" and "choice". Several versions of the tool currently exist. A full offline version in MATLAB is developed for use in a local environment. A free access prototype online version applying Hypertext Transfer Protocol (HTTP), which is the focus of this work, is also available. The online feature of this version allows a wider reach for the tool and its underlying models within the academic and industrial communities. A built-in database is included in this version, offering default parameters and properties to aid efforts especially in earlier process development and design stages. A licensed, extended online version of the tool is also currently under development to implement the full range of features and functionalities of the current offline version. The online tool can be accessed from this Uniform Resource Locator (URL): http://www.pse.t.u-tokyo.ac.jp/TECHoice/ (tested with Google Chrome Version 75.0.3770.100). Our current study focuses on presenting the online prototype version of the tool, describing: the background setup, the range of input functions, visualization of the output, and its role in the analysis of the results. The development of the tool allows potential industrial users access to the models and algorithms developed in an academic field.

2. Overview of the Tool "TECHoice"

2.1. Process and Equipment Technology Description

The manufacturing process for which the tool is developed is the sterile filling process of biopharmaceutical drug product manufacturing. Figure 1 shows a flow of typical biopharmaceutical drug product manufacturing processes with sterile filling. The configurations of full SUT, full MUT, and HYB—a common hybrid plant—are also shown in Figure 1 [15]. Sterile filling processes typically involve nine unit operations: retention, two-time-filtration, buffering, filling, and four-time transfer between unit operations. A piece of equipment is allocated to each unit operation with the exception of the filtration unit operations where two pieces of equipment, a filter housing and a filter membrane, are assigned to one operation. Therefore, eleven pieces of equipment are used in sterile filling processes. The two filter membranes and the set of filling tubes can only be resin-made, thus leaving eight pieces of equipment having two options for equipment material (resin or stainless steel). This yields a total of 256 process alternatives: SUT, MUT with stainless steel equipment wherever possible, and 254 available hybrid alternatives.

SUT, MUT, and hybrid technologies have different characteristics in terms of various aspects, as shown in Table 1. SUT requires shorter time for a change-over operation, which involves assembling and dismantling disposable equipment, but requires a larger number of operators as the operation is manually conducted. MUT, on the other hand, requires a longer time for cleaning and sterilization of the fixed stainless steel facility, i.e., clean-in-place (CIP) and sterilize-in-place (SIP) processes, using media such as water for injection and pure steam. As the operation is automated, the required number of operators is less than those required for SUT. The equipment installed for MUT requires larger investment cost and larger manufacturing area compared to SUT, which uses disposable and flexible resin-made equipment. In SUT, leachables—chemical compounds released from the resin—are a typical concern, while residue caused by cross-contamination from previous drug production, is a typical concern of MUT. Different reasons can cause manufacturing delays in SUT and MUT, affecting supply robustness. For example, delay in transportation of disposables from vendors to pharmaceutical manufacturing companies is a concern in the case of SUT, and equipment failure requiring extensive maintenance is a concern for MUT.

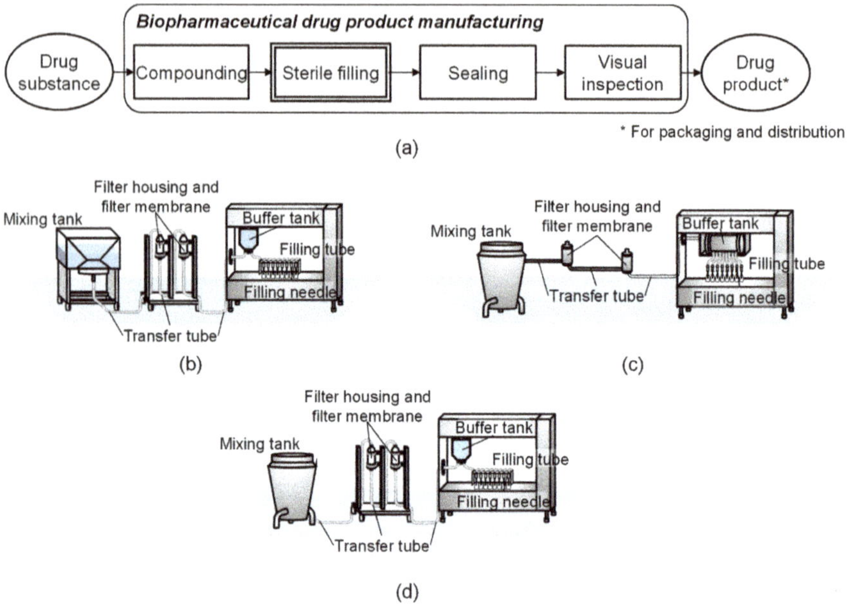

Figure 1. (**a**) Flowsheet of a typical biopharmaceutical drug product manufacturing process [15], and configuration of plants using (**b**) single-use technology (SUT) [15], (**c**) multi-use technology (MUT) [15], and (**d**) a common industrial hybrid technology option (HYB).

Table 1. Characteristics of the three equipment technologies.

Technology	Single-use (SUT)	Hybrid	Multi-use (MUT)
Required time for change-over	Short	Long	Long
Investment cost	Small	Small/Large [1]	Large
Number of operators	Large	Intermediate	Small
Manufacturing area	Small	Intermediate	Large
Usage of media	N/A	Small/Large [1]	Large
Quality issues	Leachables	Both	Residue/Cross-contamination
Supply robustness issues	Vendor dependency	Both	Equipment failure

[1] Depending on the material choice of the mixing tank.

2.2. Need for the Tool

An intensive discussion with experts from the ISPE (International Society of Pharmaceutical Engineering) Japan community of practice "PharmaPSE COP" identified the appropriate application phase, data needs, and the impediments to exposure within the community. The discussions confirmed the existence of a gap in the available tools to support decision-making in DP manufacturing, but especially highlighted the need in earlier design stages.

This tool thus aims to support the equipment technology choice, which is an important decision that affects the initial investment and manufacturing area design for pharmaceutical manufacturing companies. At such early process development phases, data are usually scarcer, and therefore, this tool offers default design options for users to best explore the possible design landscape. The tool can be used by industrial or academic research groups dedicated to investment decisions as a first indication at early decision phases. Figure 2 shows the different pharmaceutical production stages and the intended use phase for this tool.

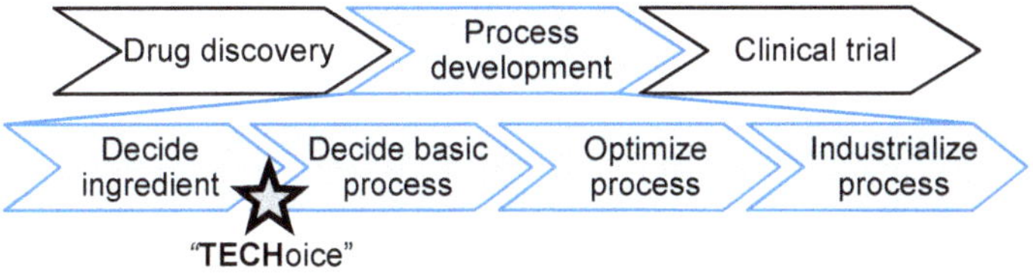

Figure 2. Intended use phase of the tool "**TECH**oice".

Another important need addressed by the tool is its convenience and accessibility due to its online feature. Several versions of the tool exist, offering different functionalities for various users. The complete version of the tool is currently written in MATLAB R2018b (The MathWorks, Inc., Natick, MA, United States). Several formats for algorithm delivery were reviewed and evaluated, such as directly using the MATLAB (.m) files or compiling them into executables (.exe) using the AppDesigner supplied by Mathworks. However, such formats may be inconvenient for industrial users since proprietary software, such as MATLAB, is sometimes unavailable or with a restricted number of licenses. Furthermore, the installation of software and contents from outside the company is normally prohibited for data security reasons. Therefore, a version of the tool was then implemented as an online web application, allowing easy access from anywhere without the need for periodical updates by users.

2.3. Key Features

Currently, two versions of the equipment technology choice decision-support tools are available: a full offline version in MATLAB and a free access prototype online version. A licensed extended online version is currently under development. The prototype online version is the focus of this work.

The online prototype version of the tool applies HTTP, which enables data communication between users and servers. User interactions on a web browser are sent as requests to a web server as shown in Figure 3. If any calculations are needed, the webserver sends a request to another calculation server, the web server receives the calculation results as a response, and the results are displayed on the user's web browser. An Apache® HTTP Server Version 2.4 (The Apache Software Foundation, Forest Hill, MD, United States) is used as the web server, and a MATLAB Production Server™ (MPS) [22] is used as the calculation server installed on a Windows Server 2016 operating system.

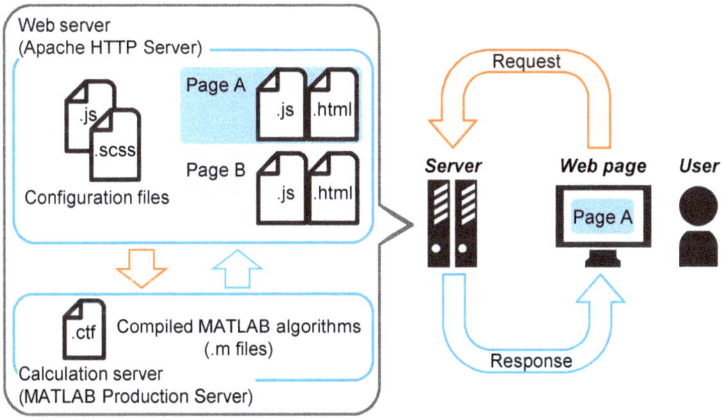

Figure 3. Structure of the online tool applying Hypertext Transfer Protocol (HTTP).

Source codes are written in TypeScript (TS), JavaScript (JS), and HyperText Markup Language (HTML) for the user interface of data input and output visualization. These codes are developed and compiled to be built on the web server. MATLAB codes have been developed based on the in-house algorithms for the offline version of the tool and compiled to be built on MPS.

2.4. Built-in Database

Generally, each user must input values for parameters in their tested cases. However, if the data is not available or the user is using the tool for exploration of alternatives rather than for a specific case, the tool then offers default values that can be used from a built-in database. The availability of default values for key model parameters is an especially useful feature for academic users who do not have access to industrial data. Data have been collected from various sources and online databases to serve as default values for model parameters. Table 2 shows the categories of database parameters with some examples and their sources. More details regarding the values used and their sources can be found in our previous work [13]. With respect to the data collected from industrial experts, average values of the data range provided by the experts have been used in this tool as default parameters, which can be freely used by anyone. Influential default parameters from the database, if selected by users, are displayed on the "evaluation target" page in the tool for confirmation by the users or target audience.

Table 2. Database parameters and their sources.

Category	Example	Sources
Physical properties of the drug product solution	Molar weight, viscosity	Online databases, e.g., ChemSpider [23], PubChem [24]
Flowsheet	Number and order of unit operations	Interviews with industrial experts, e.g., pharmaceutical manufacturing companies, equipment suppliers
Equipment configuration	Standard industrial equipment sizes	
Operating conditions	Standard change-over times, number of operators	
Price information	Prices of standard equipment and utilities	
Emission data	Resin incineration and utility consumption	Life cycle assessment (LCA) databases, e.g., JLCA-LCA database [25], LCI Database IDEA [26,27]
Properties of leachables	Saturation concentration, permitted daily exposure	Online databases, e.g., ChemSpider [23], PubChem [24]

In the online prototype version of the tool, parameters listed in Table 2 are fixed to the default values without the possibility of any user-induced changes. This option will only be available in the full licensed version. Currently, users only specify the production scenario, e.g., project lifetime, annual production volume (per plant), production mode, and number of products per year, in addition to the filling volume of the containers, e.g., vials.

2.5. Algorithms and Models

The algorithms and models implemented in the tool are based on a framework developed for generating and evaluating alternatives for sterile filling processes of drug product manufacturing [15]. The framework is composed of four decision layers as shown in Figure 4: product, flowsheet, equipment, and operating conditions. Each of the layers has some parameters with discrete options or a range of values. The order of the layers from "product" to "operating conditions" follows the decision order, i.e., the parameters in the "product" layer are determined earlier than those in the "flowsheet" layer. When all of the options and values of the parameters are specified, one process alternative is defined.

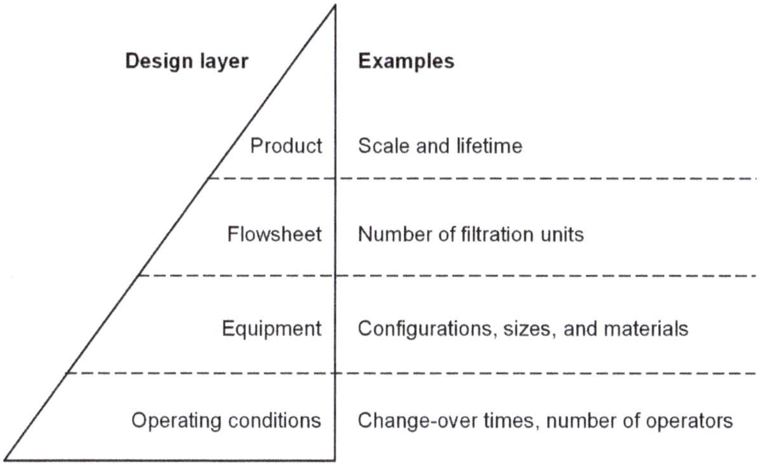

Figure 4. Decision layers within the framework for alternative generation in DP (drug product) manufacturing.

In this work, parameters, such as filling volume, lifetime, production volume, and number of products, belong to the "product" layer. Once the user specifies these parameters in the prototype version of the tool, the options or values of parameters on the "flowsheet" layer are fixed to the default setting from the database. On the "equipment" layer, equipment configuration and sizes are fixed to the default values. The full offline version of the tool varies the equipment material for each of the unit operations in the process yielding a maximum of 256 alternatives. The prototype online tool only displays results for the SUT, MUT, and most common industrial hybrid alternative (HYB), shown in Figure 1d, which has a stainless steel mixing tank and resin-made equipment everywhere else. On the "operating conditions" layer, users can define the production mode as either campaign or alternating. Campaign production implies producing the same product in back-to-back batches, whereas alternating production implies producing a different product with each batch. Operating parameters, such as batch sizes and filling time, are calculated based on the data chosen in the previous decision layers.

The generated alternatives are evaluated on the basis of four indicators. The economic aspect *Eco* [JPY] uses net present value, and the environmental aspect *Env* [kg-CO_2] uses life cycle CO_2 emissions, as indicators. The product quality indicator *PQ* [–] evaluates the impact of patient exposure to leachables and the potential risk of patient exposure to residues. The supply robustness indicator *SR* [–] describes the risk of production delays. For detailed model assumptions and equations, see our previous work [13,15]. Ultimately, the evaluated results with four indicators are aggregated to one indicator, total score *T* [–], given by:

$$T = w_{\text{economy}}\, Eco/Eco' + w_{\text{environment}}\, Env/Env' + w_{\text{supply}}\, SR/SR' + w_{\text{quality}}\, PQ \quad (1)$$

where different weighting factors: w_{economy}, $w_{\text{environment}}$, w_{supply}, and w_{quality}, are used for economic impact, environmental impact, supply robustness, and product quality, respectively. Here, *Eco'*, *Env'*, and *SR'* are the maximum values of economic impact, environmental impact, and supply robustness among evaluated 256 alternatives, respectively. *PQ* is an aggregated value of two different product quality impacts that are leachables *Lea* [–] and residue *Res* [–] given by:

$$PQ = w_{\text{leachables}}\, Lea/Lea' + w_{\text{residue}}\, Res/Res' \quad (2)$$

where $w_{\text{leachables}}$ and w_{residue} are the weighting factors of leachables and residue, respectively. The parameters *Lea'* and *Res'* are the maximum values of leachables and residue impacts among evaluated 256 alternatives, respectively.

2.6. Tool Architecture and Key Input Parameters

The prototype tool evaluates SUT, MUT, and HYB using multi-objective indicators. Figure 5 shows the architecture of the tool "**TECH**oice" which comprises four web pages.

1. "Start page" as a front page to show an explanation of the tool and the data management policy;
2. "Data input" as a data entry form of, e.g., product specifications and production scenario;
3. "Evaluation target" as a value confirmation page of physical properties of the drug product solution, the choice of flowsheet, operating conditions, equipment configuration, sizes, and material;
4. "Evaluation results" as a page showing the overall results, category results breakdown, and sensitivity analysis.

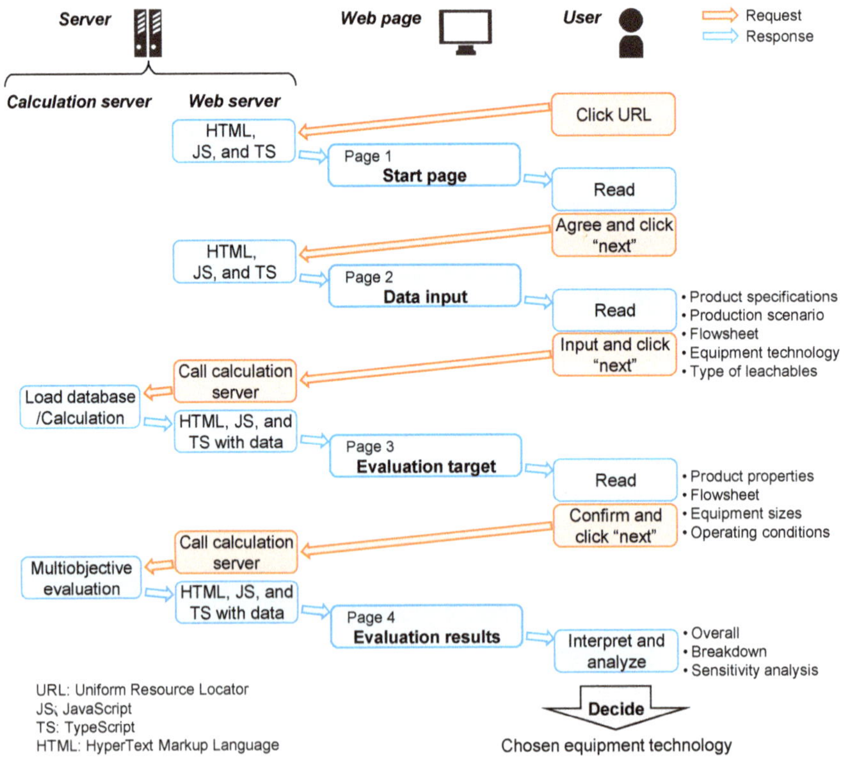

Figure 5. Architecture of "**TECH**oice" with flows of requests and responses over four web pages.

The main input parameters of the tool are product specifications, such as filling volume, and production scenario, e.g., project lifetime, annual production volume, production mode, and number of products per year. The possible ranges of the input parameters are shown in Table 3. According to the user input data on the second page, the physical properties of the drug product solution and the flowsheet are loaded from the built-in database, and equipment sizes and operating conditions, such as batch size and filling time, are calculated. After users confirm the results on the third page, multi-objective evaluation is conducted with four indicators: economic and environmental impacts as well as product quality and supply robustness, where price data, emission data, or properties of leachables are loaded from the built-in database for the calculation. Overall results and a breakdown of the results for each of the four indicators are shown for both SUT and MUT on the last page. The current version of the tool offers different components of the economic and environmental results (disposables, labor, utilities, and investment). However, despite individual equipment costs being a part of the calculation, their values are not accessible in the current version. This feature will only be available in the licensed online version. The total score, which integrates values of all the four indicators, is shown for SUT, MUT, and HYB using weighting factors that reflect the indicator priorities of the user. The total score determines the final technology choice where a smaller score value indicates a better technology option. The results are sometimes sensitive enough to the indicator weighting factors to change the choice of the preferred equipment technology. Therefore, the final page has a function that can be used to display results of the sensitivity analysis to the weighting factors. The default setting is equal weighting among the four indicators. Discrete options can be chosen to test the sensitivity in some cases, such as "economy first" which allocates a minimum weighting factor of 0.1 to all indicators other than the economic aspect (allocated 0.7). Similarly, cases of "environment first", "supply first", "quality first", and "supply and quality first" are also evaluated.

Table 3. Possible range of input parameters.

Category	Parameter	Unit	Available Ranges or Values
Product specifications	Product type Filling volume Target disease	– L/vial –	Default >0 Default
Production scenario	Project lifetime Annual production volume Production mode Number of product per year	y L/y – product/y	$0 < x \leq 20$ >0 {Campaign, Alternating} ≥1
Flowsheet		–	Default
Equipment technology	Equipment technology Equipment material for HYB	– –	{SUT, MUT, HYB} {Stainless steel, Resin}
Evaluation	Chemical compounds to evaluate as leachables	–	Default

3. Case Study

We conducted a case study to demonstrate the use of the tool with four different production scenarios. The following sections outline the details and visualization of "**TECH**oice" from page 1, "start page"; page 2, "data input"; page 3, "evaluation target"; to page 4, "evaluation results". Screenshots (as of 11 July 2019) of all four pages are shown in the Appendix A.

3.1. Start Page

This is the front page depicting an explanation of the tool, such as purpose, target, and key input parameters, and the data management policy as shown in Figure A1. The users click the "next" button after agreeing with the terms.

3.2. Data Input

The purpose of the page is for users to insert their input parameters. There are five sections where users are required to fill in values or choose options: "Define your product", "Define your production scenario", "Select your flowsheet", "Select your options of equipment technology", "Fill in the type of chemical compounds you want to evaluate as leachables" as shown in Figure A2.

In this case study, input parameters of four production scenarios were demonstrated. Table 4 shows the list of input values for each scenario. The tested scenarios are defined as follows:

- Scenario A: large-scale in campaign production mode with minimal change-over;
- Scenario B: large scale in alternating production mode with maximal change-over;
- Scenario C: small-scale in campaign production mode with minimal change-over;
- Scenario D: small-scale in alternating production mode with maximal change-over.

Table 4. Input parameters of each scenario.

		Scenario			
Parameter	**Unit**	**A**	**B**	**C**	**D**
Project lifetime	y	10	10	10	10
Annual production amount (per plant)	L/y	150,000	150,000	10,000	10,000
Production mode	–	Campaign	Alternating	Campaign	Alternating
Number of product per year	product/y	2	10	2	10

In this case study, minimal and maximal change-overs were taken as two and ten products, respectively.

In the product definition section, a monoclonal antibody for bowel cancer with the filling volume of 0.015 L/vial was assumed. The default drug product solution properties were taken to be the same as water in the prototype version. The target disease is used to estimate the potential patient demand size, which is used to evaluate the supply robustness indicator in the tool. The default flowsheet was assumed to have the same process as explained in Section 2.1. Three types of equipment technologies were considered: SUT, MUT, and HYB. The prototype version of the tool takes stearic acid as the default compound leaching from resin-made equipment since it is a common example of leachables.

3.3. Evaluation Target

The purpose of this page is for the users to confirm their input values, values from database, default assumptions, and intermediate calculation results. This page is composed of five sections: "Product data", "Flowsheet data", "Equipment data", "Operating conditions", and "Evaluation parameters".

Table 5 shows some of the intermediate calculation results of batch size, filling time, and annual number of batches, in addition to default values from the database, such as required time for change-over and number of operators. Since alternating production features more time-intensive product-to-product changeover operations, it can therefore accommodate a smaller number of annual batches (Scenarios B and D) compared to campaign production modes (Scenarios A and C). The scenarios with larger annual production volume (Scenarios A and B) were assigned larger batch sizes to fit into the fixed annual working time. The filling time was proportional to the batch size due to the fixed pumping speed. The number of batches calculated for Scenario C was higher than expected for a realistic industrial production case. In this prototype version, however, no error messages will be displayed for such a case. The full version will display a warning message.

3.4. Evaluation Results

The purpose of this page is to visualize the multi-objective evaluation results for interpretation and choice of technologies. The page has five sections: "Overall", "Result for SUT", "Result for MUT", "Total score", and "Sensitivity analysis".

Table 5. Parameters of the operating conditions decision layer of different (**a**) scenarios (intermediate calculation results) and (**b**) technologies (default values from the database).

(a)		Scenario			
Parameter	Unit	A	B	C	D
Batch size	L	656	1725	26.8	70.9
Filling time	h	3.04	8.00	0.124	0.328
Annual number of batches	/y	229	87	373	141
(b)		**Technology**			
Parameter	**Unit**	**SUT**	**HYB**	**MUT**	
Required time for batch-to-batch change-over operation	h	2	4.5	4.5	
Required time for product-to-product change-over operation [1]	h	2	7.5	7.5	
Number of operators	–	5	4	2	

[1] Change-over operations in SUT (single-use technology) do not differentiate batch-to-batch and product-to-product change-over operations, whereas HYB (a common industrial hybrid technology option) and MUT (multi-use technology) have different change-over types depending on the operation.

3.4.1. Overall/ Result for SUT/ Result for MUT

In Section 1, indicator values for the economic and environmental impacts, product quality, and supply robustness aspects are shown in the form of a table. The next two sections show the breakdown of the economic and environmental impacts of SUT and MUT in pie charts in terms of the contribution of disposables, labor, utilities, and investment costs.

The breakdown of economic and environmental impacts of Scenario A (large-scale in campaign production mode with two products) and D (small-scale in alternating production mode with ten products) are shown in Table 6. The breakdown of the economic impact results showed similar common trends between scenarios A and D. SUT incurs a large cost for purchasing disposable equipment and minimal investment costs due to the lack of fixed stainless steel equipment. The overall results showed that the economic impact of SUT was larger than that of MUT in Scenario A, and the impact of MUT was larger than that of SUT in Scenario D. Scenario A had a larger annual number of batches which led to more change-over in disposable equipment, causing the larger cost of SUT. Labor cost in SUT was larger for Scenario A due to the larger number of operators compared to MUT. Scenario D, however, featured frequent product-to-product change-over operations in the alternating production mode. Given that the time for product-to-product change-over of MUT is longer than that of SUT, operator working hours were longer in this scenario leading to higher labor costs for MUT.

The breakdown of the environmental results showed similar common trends between Scenarios A and D. The impact from disposables was larger in SUT compared to MUT, where the contribution of utility consumption was large for both SUT and MUT. The environmental impact of utility consumption stems from the energy required for heating, ventilation, and air conditioning (HVAC) to keep the manufacturing area clean. In the environmental impact calculation, manufacturing area and time are the key parameters affecting process utility consumption. In the default setting, MUT with fixed piping is assumed to have double the size of manufacturing area compared to SUT with flexible tube, causing the impact of utilities in MUT to be larger than that in SUT.

Table 6. Breakdown of economic and environmental impacts of Scenarios (**a**) A and (**b**) D.

(a)	**Scenario A(Large-scale in campaign production mode with two products)**			
	Economic impact (JPY)		**Environmental impact (kg-CO_2)**	
	SUT	**MUT**	**SUT**	**MUT**
Disposables	5.23×10^8	9.67×10^7	45.3	4.74
Labor	2.74×10^8	1.65×10^8	–	–
Utilities	5.48×10^7	1.67×10^8	76.4	254
Investment	9.00×10^5	9.73×10^7	–	–
Overall	8.53×10^8	5.27×10^8	122	259
(b)	**Scenario D** (Small-scale in alternating production mode with ten products)			
	Economic impact (JPY)		**Environmental impact (kg-CO_2)**	
	SUT	**MUT**	**SUT**	**MUT**
Disposables	2.79×10^8	5.96×10^7	38.4	4.74
Labor	7.81×10^7	1.65×10^8	–	–
Utilities	1.56×10^7	1.68×10^8	35.3	440
Investment	4.99×10^5	9.14×10^7	–	–
Overall	3.73×10^8	4.85×10^8	73.7	444

3.4.2. Total Score

In Section 4, the total score, an aggregated value of all the four evaluation results, is shown in a stacked bar chart for the three evaluated alternatives: SUT, MUT, and HYB. Figure 6 shows the results of the total score of the four tested scenarios. The alternative with the smallest total score is to be chosen. In the case of alternating production of ten products (B, D), SUT was chosen as the best alternative regardless of the annual production volume. In the case of campaign production with two products (A, C), MUT was the best alternative regardless of the annual production volume. The difference in the total score between SUT and MUT in Scenarios A and C is too small to make a decision with confidence. Individual indicators, however, show different profiles between SUT and MUT. In this case, the weighting factors of the different indicators play a significant role in the final decisions, and a more in-depth analysis of the individual indicators' results is required. Among the four tested scenarios, the decision with the highest certainty was the choice of SUT in Scenario D as indicated by a difference in the total score of SUT compared to the others. To finalize the decision, the values of the weighting factors were varied for this scenario to test the effect of the variance on the conclusion.

3.4.3. Sensitivity Analysis

The sensitivity analysis function is embedded below the stacked bar charts of total score, where users can choose six different combinations of weighting factors depending on the priority of aspects: equal weighting, economy first, environment first, supply first, quality first, supply and quality first. The bar charts showing the results of the total score change according to the change with the total score calculated from the selected combination of weighting factors.

Figure 7 shows the results of the sensitivity analysis to the impact of weighting factors on the assessment results for Scenario D (small-scale in alternating production mode with ten products). Weighting factors of the four indicators were discretely changed. The default combination used in the assessment is equal weighting factors ($w = 0.25$) for all indicators (same as the results already shown in Figure 6). Equal weighting of the indicators showed SUT as the best alternative for Scenario D. When supply robustness was prioritized, the total score of MUT was the smallest due to the lower probability of delays due to equipment failures than delays due to supply failure of the disposable equipment. HYB was the best when product quality was prioritized. The impact of product quality is the aggregated effect of leachables and residues, where the effect of leachables was the largest in SUT, second largest in HYB, and the smallest in MUT. On the other hand, the effect of residue was the largest in MUT, second largest in HYB, and the smallest in SUT. When the product quality impact was calculated using Equation (2) with the same weighting factors for leachables and residue, the impact of HYB became the smallest among the three alternatives. The HYB alternative has a smaller exposure to leachables compared to SUT since it employs a stainless steel mixing tank, which is the equipment with the largest area and residence time, and thus the highest contribution to leachable concentration in the system. Residue concentration is assumed to be a function of only the contact surface area rather than the residence time. Since the mixing tank's area in this scenario is smaller than all other equipment combined, the expected residue concentration is therefore smaller in the HYB alternative compared to MUT. The combined effect of leachables and residue was also smaller compared to either SUT or MUT.

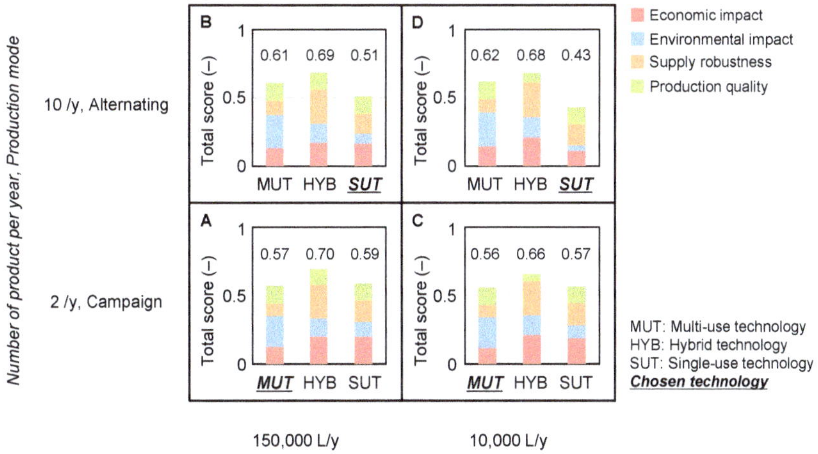

Figure 6. Landscape results of the total score for different production scenarios of (**A**) large-scale in campaign production mode with minimal change-over, (**B**) large scale in alternating production mode with maximal change-over, (**C**) small-scale in campaign production mode with minimal change-over, and (**D**) small-scale in alternating production mode with maximal change-over.

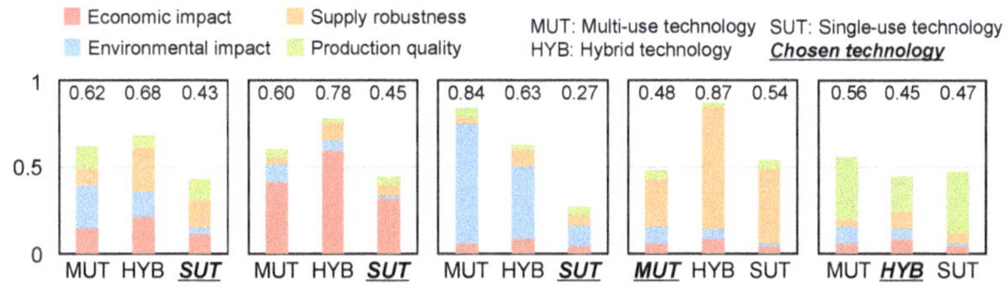

	Equal weighting	Economy first	Environment first	Supply first	Quality first
$w_{economy}$	0.25	0.7	0.1	0.1	0.1
$w_{environment}$	0.25	0.1	0.7	0.1	0.1
w_{supply}	0.25	0.1	0.1	0.7	0.1
$w_{quality}$	0.25	0.1	0.1	0.1	0.7
	(a)	(b)	(c)	(d)	(e)

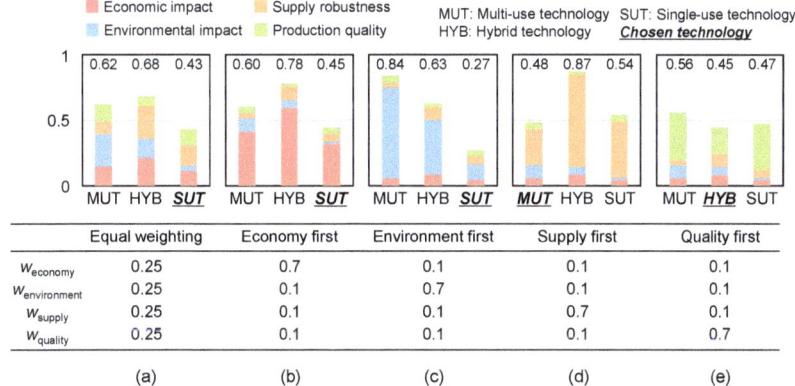

Figure 7. Sensitivity analysis of total score for Scenario D by (**a**) applying equal weighting for the four indicators, (**b**) prioritizing economic impact, (**c**) prioritizing environmental impact, (**d**) prioritizing supply robustness, and (**e**) prioritizing production quality.

4. Discussion

The equipment technology choice among SUT, MUT, and HYB can thus be made based on the analysis presented in the case study for each scenario. Users can form a comprehensive picture of the possible range of assessment results. The advantages of the tool include flexibility in changing input parameters as well as visualization of the multi-objective evaluation and sensitivity analysis results. The tool with these benefits establishes a basis for a platform that connects academically developed models and algorithms to users in the industrial community with real production decisions.

The prototype tool can still be extended to a full online version offering more functionalities, alternatives, and analysis features. For example, flexibility of input parameters could be upgraded. Only "default" values are allowed for "product type" and "flowsheet" in the current version. In the extended version, more "customized" choices will be allowed. In addition, in the prototype version no error messages are shown regardless of values or options that users selected, e.g., negative values of input parameters, such as project lifetime or annual production volume. Warnings are not given for unrealistic production conditions, e.g., unrealistic batch numbers or batch sizes. Disposable mixing tanks higher than 2000 L are not available, however, SUT options with higher batch sizes would not get a warning message.

Output visualization will be improved to display results for other hybrid alternatives and to show a more detailed breakdown of indicator results. In the current version, only results of SUT, MUT, and an empirical hybrid alternative can be seen in full detail. More extensive sensitivity analysis can help users identify critical process parameters of different design stages. The uncertainty analysis concerning the influence of various model parameters and their ranges on the results will be the focus of our next publication, which aims to parameterize and landscape at different design phases. Technical updates can also be expected, e.g., for the user interface to visualize the results of the extended sensitivity analysis, or security updates for the input data.

Author Contributions: Conceptualization, H.S. (Haruku Shirahata), Y.S., S.H., S.B., and H.S. (Hirokazu Sugiyama); methodology, H.S. (Haruku Shirahata), S.B., and H.S. (Hirokazu Sugiyama); software, H.S. (Haruku Shirahata), Y.S., and S.H.; investigation, H.S. (Haruku Shirahata), Y.S., and S.H.; data curation, H.S. (Haruku Shirahata); writing—original draft preparation, H.S. (Haruku Shirahata); writing—review and editing, S.B. and H.S. (Hirokazu Sugiyama); supervision, H.S. (Hirokazu Sugiyama); project administration, H.S. (Hirokazu Sugiyama); funding acquisition, H.S. (Hirokazu Sugiyama).

Funding: This research was funded by Grant-in-Aid for Young Scientists (A) No. 17H04964 and Research Fellow (DC2) No. 18J13892 from the Japan Society for the Promotion of Science.

Acknowledgments: The authors acknowledge Rizki Darmawan from The University of Tokyo, industrial experts from the International Society of Pharmaceutical Engineering (ISPE) Japan community of practice "PharmaPSE COP", especially Akito Daiba from Dassault Systèmes K.K., and Kojiro Saito from MathWorks.

Conflicts of Interest: The authors declare no conflict of interest.

Appendix A

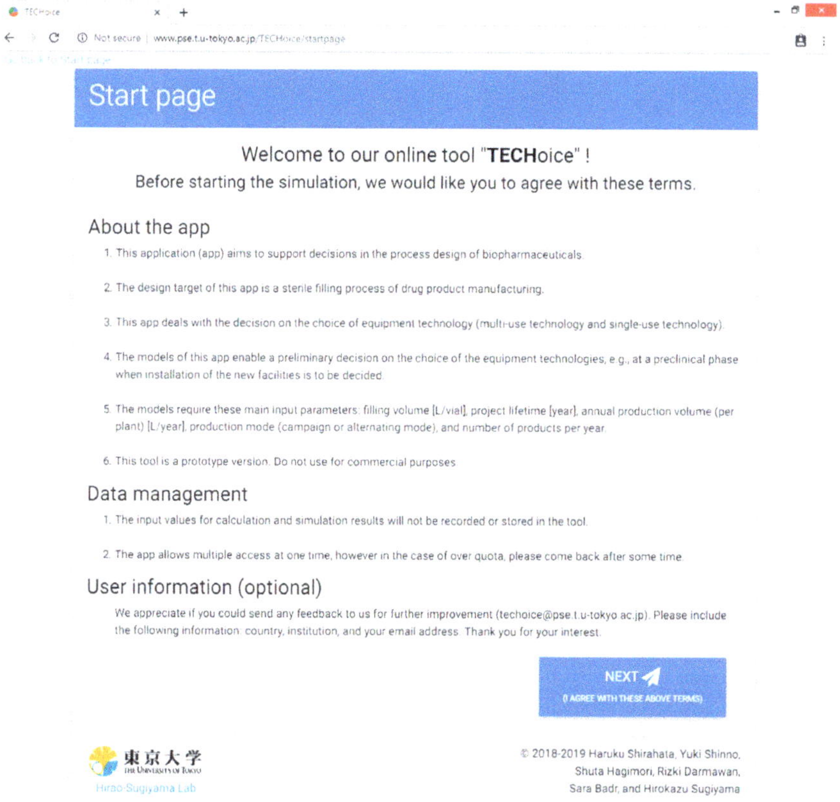

Figure A1. Screenshot of the first page (Start page) of "**TECH**oice".

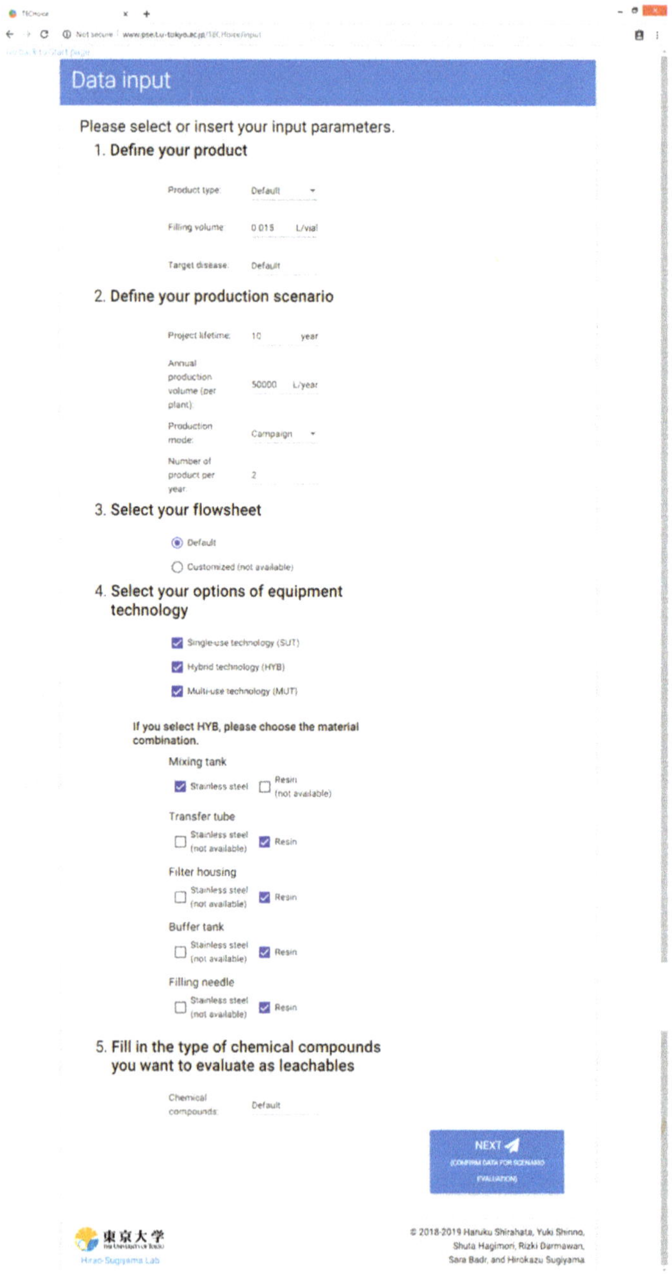

Figure A2. Screenshot of the second page (Data input) of "**TECH**oice".

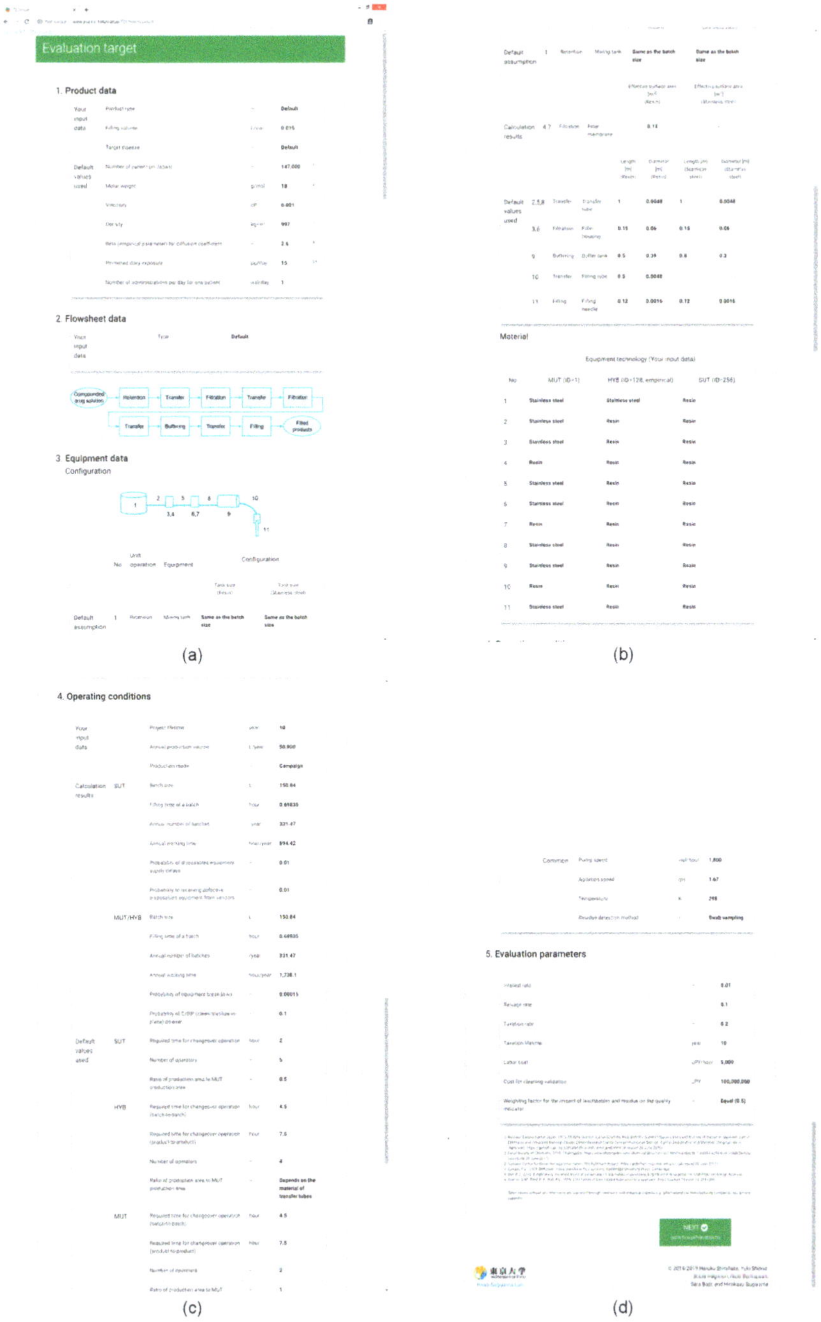

Figure A3. (**a–d**) Screenshot of the third page (Evaluation target) of "**TECH**oice".

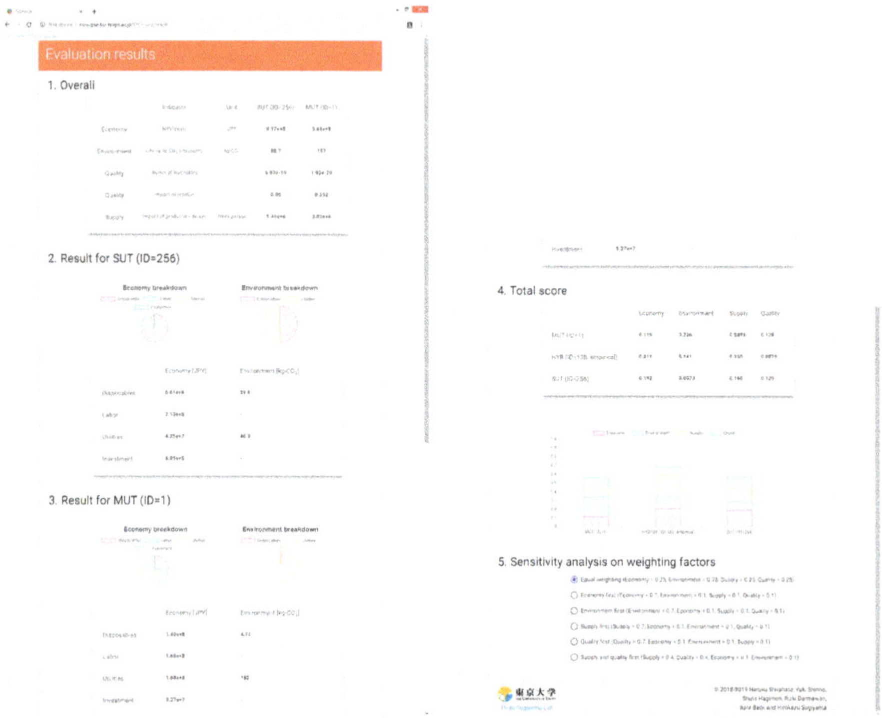

(a) (b)

Figure A4. (**a**,**b**) Screenshot of the fourth page (Evaluation results) of "**TECH**oice".

References

1. U.S. Food and Drug Administration (FDA) CFR—Code of Federal Regulations Title 21. Available online: https://www.accessdata.fda.gov/scripts/cdrh/cfdocs/cfcfr/CFRSearch.cfm?fr=211.67 (accessed on 7 June 2019).
2. Lopes, A.G. Single-use in the biopharmaceutical industry: A review of current technology impact, challenges and limitations. *Food Bioprod. Process.* **2015**, *93*, 98–114. [CrossRef]
3. Klutz, S.; Magnus, J.; Lobedann, M.; Schwan, P.; Maiser, B.; Niklas, J.; Temming, M.; Schembecker, G. Developing the biofacility of the future based on continuous processing and single-use technology. *J. Biotechnol.* **2015**, *213*, 120–130. [CrossRef] [PubMed]
4. Ierapetritou, M.; Muzzio, F.; Reklaitis, G. Perspectives on the continuous manufacturing of powder-based pharmaceutical processes. *AIChE J.* **2016**, *62*, 1846–1862. [CrossRef]
5. Karst, D.J.; Steinebach, F.; Morbidelli, M. Continuous integrated manufacturing of therapeutic proteins. *Curr. Opin. Biotechnol.* **2018**, *53*, 76–84. [CrossRef] [PubMed]
6. Metta, N.; Ghijs, M.; Schäfer, E.; Kumar, A.; Cappuyns, P.; Assche, I.V.; Singh, R.; Ramachandran, R.; Beer, T.D.; Ierapetritou, M.; et al. Dynamic Flowsheet Model Development and Sensitivity Analysis of a Continuous Pharmaceutical Tablet Manufacturing Process Using the Wet Granulation Route. *Processes* **2019**, *7*, 234. [CrossRef]
7. Lee, J.C.; Chang, H.N.; Oh, D.J. Recombinant Antibody Production by Perfusion Cultures of rCHO Cells in a Depth Filter Perfusion System. *Biotechnol. Prog.* **2005**, *21*, 134–139. [CrossRef] [PubMed]
8. Sandstrom, C. Disposable vs. Traditional Equipment- A facility-Wide View. *Chem. Eng. Prog.* **2009**, *105*, 30–35.
9. Monge, M.; Sinclair, A. Disposables Cost Contributions: A Sensitivity Analysis. *BioPharm Int.* **2009**, *22*.

10. Pietrzykowski, M.; Flanagan, W.; Pizzi, V.; Brown, A.; Sinclair, A.; Monge, M. An environmental life cycle assessment comparison of single-use and conventional process technology for the production of monoclonal antibodies. *J. Clean. Prod.* **2013**, *41*, 150–162. [CrossRef]
11. Rawlings, B.; Pora, H. Environmental impact of single-use and reusable bioprocess systems. *Bioprocess Int.* **2009**, *7*, 18–25.
12. Shirahata, H.; Hirao, M.; Sugiyama, H. Decision-Support Method for the Choice Between Single-Use and Multi-Use Technologies in Sterile Drug Product Manufacturing. *J. Pharm. Innov.* **2017**, *12*, 1–13. [CrossRef]
13. Shirahata, H.; Hirao, M.; Sugiyama, H. Multiobjective decision-support tools for the choice between single-use and multi-use technologies in sterile filling of biopharmaceuticals. *Comput. Chem. Eng.* **2019**, *122*, 114–128. [CrossRef]
14. Farid, S.S.; Washbrook, J.; Titchener-Hooker, N.J. Decision-support tool for assessing biomanufacturing strategies under uncertainty: Stainless steel versus disposable equipment for clinical trial material preparation. *Biotechnol. Prog.* **2005**, *21*, 486–497. [CrossRef] [PubMed]
15. Shirahata, H.; Badr, S.; Dakessian, S.; Sugiyama, H. Alternative generation and multiobjective evaluation using a design framework: Case study on sterile filling processes of biopharmaceuticals. *Comput. Chem. Eng.* **2019**, *123*, 286–299. [CrossRef]
16. Aspen Technology, Inc. Aspen Plus. 2019. Available online: https://www.aspentech.com/en/products/engineering/aspen-plus (accessed on 14 June 2019).
17. Aspen Technology, Inc. Aspen HYSYS. 2019. Available online: https://www.aspentech.com/en/products/engineering/aspen-hysys (accessed on 14 June 2019).
18. Aspen Technology, Inc. Aspen Batch Process Developer. 2019. Available online: https://www.aspentech.com/products/engineering/aspen-batch-process-developer/ (accessed on 14 June 2019).
19. Farid, S.S.; Washbrook, J.; Titchener-Hooker, N.J. Modelling biopharmaceutical manufacture: Design and implementation of SimBiopharma. *Comput. Chem. Eng.* **2007**, *31*, 1141–1158. [CrossRef]
20. Biopharm Services Limited. BioSolve Professional ver. 7.0.6.3. 2018. Available online: https://biopharmservices.com/software/ (accessed on 14 June 2019).
21. OUAT! HakoBio. 2019. Available online: http://www.hakobio.com/ (accessed on 14 June 2019).
22. The MathWorks, Inc. MATLAB Production Server. 2019. Available online: https://jp.mathworks.com/products/matlab-production-server.html (accessed on 14 June 2019).
23. Royal Society of Chemistry, ChemSpider. 2015. Available online: http://www.chemspider.com/ (accessed on 14 June 2019).
24. National Center for Biotechnology Information, The PubChem Project. Available online: https://pubchem.ncbi.nlm.nih.gov/ (accessed on 14 June 2019).
25. Life Cycle Assessment Society of Japan. *JLCA-LCA Database*, 4th ed.; Chiyoda-ku, Tokyo, Japan, 2014.
26. Japan Environmental Management Association for Industry (JEMAI). *LCA System MiLCA (original title in Japanese), ver. 1.1.*; Chiyoda-ku, Tokyo, Japan, 2012.
27. National Institute of Advanced Industrial Science and Technology and JEMAI. *LCI Database IDEA (original title in Japanese), ver. 1.1.*; Chiyoda-ku, Tokyo, Japan, 2012.

processes

Concept Paper

Show Me the Money! Process Modeling in Pharma from the Investor's Point of View

Christos Varsakelis *, Sandrine Dessoy, Moritz von Stosch and Alexander Pysik

Technical Research & Development, GSK Biologicals, 1330 Rixensart, Belgium
* Correspondence: christos.x.varsakelis@gsk.com

Received: 31 July 2019; Accepted: 2 September 2019; Published: 4 September 2019

Abstract: Process modeling in pharma is gradually gaining momentum in process development but budget restrictions are growing. We first examine whether and how current practices rationalize within a decision process framework with a fictitious investor facing a decision problem subject to incomplete information. We then develop an algorithmic procedure for investment evaluation on both monetary and diffusion-of-innovation fronts. Our methodology builds upon discounted cash flow analysis and Bayesian inference and utilizes the Rogers diffusion of innovation paradigm for computing lower expected returns. We also introduce a set of intangible metrics for quantifying the level of diffusion of process modeling within an organization.

Keywords: process modeling; return on investment; diffusion of innovation

1. Introduction

Modeling and simulation (M&S) refers to the R&D (Research & Development) methodology where mathematical equations (models) are solved numerically or analytically (via simulation) for the description of physical systems. Such a generic definition captures all different types of representations of physical systems: Mechanistic, empirical and hybrid. However, in this paper we only focus on mechanistic and/or hybrid models. Modeling and simulation (M&S) has been gradually adopted by different industries for the understanding, investigation, optimization and diagnostics of existing and future processing technologies since the 1960s giving rise to what is commonly referred to as process modeling.

The pharmaceutical industry constitutes an interesting case. On the one hand, computational chemistry has long been an indispensable tool in drug discovery and, nowadays, in silico drug discovery, it is spearheading future developments. On the other hand, the pharmaceutical industry is among the last ones to join the party since process modeling has only been sporadically utilized despite advocates preaching for the contrary [1,2]. This thought-provoking conundrum has not gone-by unnoticed and there is a wealth of efforts devoted to its study [3–6]. A synthesis of the results has revealed several factors with the most recurring ones being:

(i) Keeping science out of processing. This manifests itself through the continuous and oftentimes erroneous belief that (a) the complexity of the processes is too high and (b) the maturity of M&S is too low for the production of fruitful results. This line of thought has been perpetuating though some recent efforts that hint that blending science-based solutions with engineering approaches is growing momentum [7]. Moreover, and perhaps more importantly, there is a growing volume of research efforts (i) corroborating both the pertinence and the efficacy of M&S on both upstream and downstream [8–10], (ii) offering holistic and industrial-friendly frameworks [11] and (iii) focusing on even the most novel processing techniques [12].

(ii) Lack of regulatory frameworks. M&S has been notably absent from regulatory frameworks. However, recent publications [13], betoken that such ideas are cultivating.

(iii) Domination of empirical/statistical modeling. Processing in pharma has partnered very well with statistics. Progressively, statistical modeling has been integrated in the core of R&D methodologies. Proposing alternative methodologies will undoubtedly be subject to "appeal-to-tradition" reactions.
(iv) Emphasis on drug discovery: From an investment-risk portfolio management point of view, investments in drug/vaccine discovery are more promising than those in process development/understanding. Consequently, only the bare minimum has been done to get the processes economically viable. Even so, investment-related decision making has been relevant; rationally choosing, for example, between batch and continuous processing has attracted considerable attention [14].
(v) Shortage of in-house M&S expertise. Accommodation of M&S components that are relatively new and evolving requires dedicated FTEs (Full Time Equivalent) and building up competencies. In the absence of an interest towards M&S, such internal expertise is cumbersome to be built and updated. Consequently, new concepts or breakthroughs, are difficult to detect, digest and eventually implement.

Nonetheless, process development has started to utilize elements of M&S oftentimes in a systematic fashion as part of an organizations vision for digitization [15,16] and the accommodation of the quality by design paradigm [13]. Moreover, given the persistently disappointing figures on return on investment in pharma and the gloomier predictions [17], and the ever-growing development cost and risk [18,19], acceleration of development has become a key management target [20], and here M&S is expected to yield significant results. However, in such an environment which requires a stricter scrutiny of investments, M&S teams should be prepared to address questions on the business engineering front. Put simply, the diplomatic immunity granted to M&S has been relinquished.

Designing a business case for M&S is an arduous task because, although M&S costs are straightforward to compute, outcomes of M&S exercises are laborious to quantify. What complicates matters more, is the qualitative nature of such outcomes that render relevant efforts even more challenging. Importantly, the described challenges are not confined within pharma but invariably extend to other industries which explains the dearth of relevant studies in the literature.

To the best of our knowledge, the first organization that systematically investigated the business case of M&S and openly archived it is the U.S. Department of Defense. In a series of landmark publications [21–25], the authors have investigated the evaluation of M&S returns and presented real case studies. A handful of subsequent studies have adjusted these findings though predominately in a qualitative direction. With respect to pharma, in particular, we are only acquainted with the study of [26] where the authors examine the effects of M&S in drug development and time to market and find a positive correlation in turn backed by the presentation of NPV (Net Present Value) values.

The objective of this paper is to examine process modeling in pharma from an investor's point of view and bring forward an algorithmic methodology that allows for the development of detailed business studies. Our methodology is endowed with both tangible and intangible metrics to provide for a holistic approach to the problem in hand. On the tangible front, we examine M&S under the prism of discounted cash flow analysis. As in regard to intangible metrics, our analysis draws from and builds upon the earlier studies of [21–25] but incorporates them into a diffusion-of-innovation paradigm based on the Rogers innovation curve [27].

2. State of the Art in Decision Making

Pharmaceutical corporations have already invested non-negligible amounts of capital for building up M&S competency and internal capabilities. We model the current situation and use this framework as a vehicle to optimize current practices. Let Mr. X be the budget owner of the R&D organization within a pharmaceutical corporation. Mr. X is endowed with an annual budget of M$ (dollars) that covers for both recurring (e.g., salary) and one-time costs.

At some point, Mr. X is visited by a group of managers and/or scientists, henceforth referred to as "the Group", who propose to form a M&S team focusing on processing. They require an upfront investment of $M_1 < M$ dollars per year plus $M_2 < M$ dollars for one-time costs. In support of their request, the Group typically offers four anecdotal or poorly tractable quantitative arguments: (a) Reduction of design/investigation time, (b) enhancement or replacement of real-life tests, (c) circumvention of limitations of funding, (d) insight into issues unapproachable by alternatives. Intuitively one expects that these four arguments are to a certain extent true. However, whether the aggregate effect remains positive or there is a fallacy of composition has yet to be robustly demonstrated. In layman terms, M&S can positively impact practices in processing but at what cost.

In cash-flow terms, our Group argues that the evolution of cash-flow will initially be negative, as expected, but it will gradually shift upwards and eventually become positive as pedantically shown in Figure 1a; In the absence of relevant data, figures in the present conceptual paper are rather ad-hoc representing the accumulated empirical knowledge and industrial experience of the authors. Manifestly, they offer a high-order qualitative illustration of the underlying trends and they should be interpreted as such by the reader. Nonetheless, the same figure has superimposed an alternative scenario where the cash-flow remains negative for a prolonged period. Given that penetration of a new technology/method typically follows a Rogers S-shape curve [27] depicted in Figure 1b, the plausibility of this scenario should not be ignored.

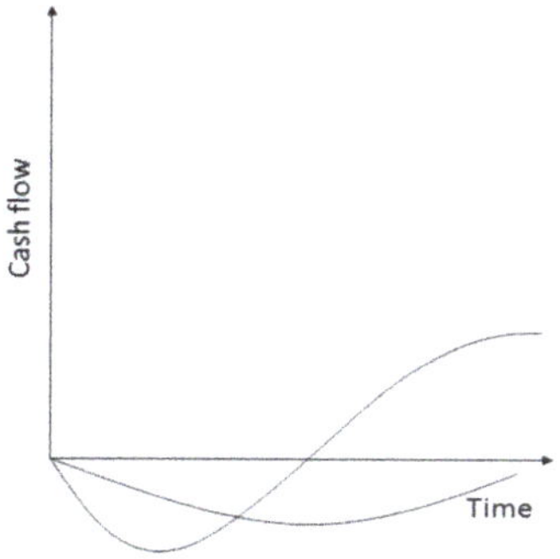

(a)

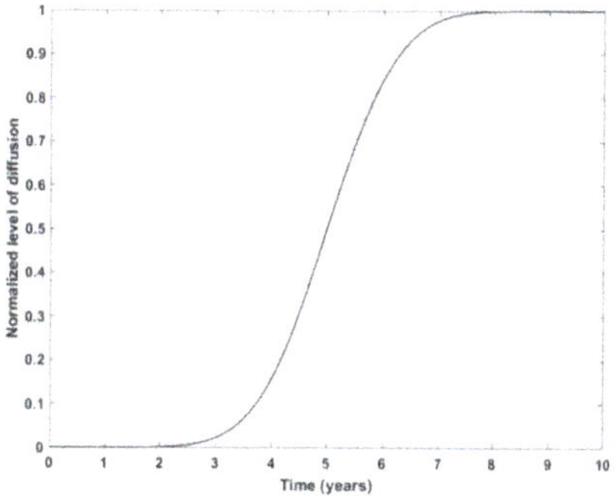

(b)

Figure 1. (**a**) Cash flow of modeling and simulation (M&S) as the time-history of profit, (**b**) A theoretical Rogers curve for diffusion/penetration of innovation.

How should our investor react? Mr. X faces an interesting decision problem. Under the assumption that Mr. X is a rational agent, the whole decision process can be modeled quite nicely, though a detailed modeling framework of this problem is quite subtle. For the sake of simplicity, herein, we sketch the basic ideas. In this respect, Mr. X is conditioning decisions on the outcomes of the following profit maximization problem:

$$max\ \pi = (q - wL) \tag{1}$$

where π, q, w, L stand for profit, production units in dollars, production units here is more broadly interpreted, cost per employee, and number of employees. We can generalize this to include physical capital and/or time but doing so increases complexity without further clarifying the picture. Mr. X considers that a simplified Cobb–Douglas production function adequately describes the relation between production units, physical capital and labor so that;

$$q = A\, L^a \tag{2}$$

with A, a constants determining productivity. Note that, typically, the Cobb–Douglas functions has a component related to cost of capital but we have neglected this here. The above is a classical paradigm that can be solved by Mr. X analytically. Now, the Group claims that the π that Mr. X has computed is not optimal. They point towards the existence of another group of employees (M&S experts) who can drive profits upwards. When Mr. X asks the Group to quantify their argument they posit that M&S experts follow a different Cobb–Douglas production function:

$$q_{MS} = e^{b_1} L^{b_2} \tag{3}$$

where b_1 and b_2 are random variables which reflects the uncertainty that even the Group has with respect to the quantification of benefits. With this information at hand, Mr. X can actually advance and solve the corresponding model. However, the accuracy of the predictions depends on the properties of b_1, b_2 and, more importantly, on whether these properties are known. Provocatively, a risk averse or even risk neutral inventive should reject this proposal as non-tractable!

Nevertheless, we have multiple examples where our investor Mr. X succumbs to the demand of the Group and grants the investment. A possible escape route may be found if we postulate that Mr. X is not only rational but also informed in the following sense. In the absence of M&S, employees can use a set of skills/prior knowledge S for the execution of their tasks. Furthermore, the net revenue per employee in the organization R is p \$/employee. Our investor has performed a comparative analysis with competitors and concludes that p does not reflect the true potential of the organization and that there is room for improvement. Mr. X goes a step further and theorizes that $R = R(S)$ with $R(S)' \geq 0$ and $R(S)'' \leq 0$. Then, being acquainted with the state of the art in M&S, X makes the informed decision that augmenting S with the competencies provided by M&S will result in an increase of $R(S)$ though it is not possible to predict the precise payoffs.

We now focus on the development and understanding of best practices. Strictly speaking, M&S (as well as experiments and statistical models) acts as an evaluation mechanism. For instance, an M&S exercise provides an insight to a phenomenon and as such empowers stakeholders to make informed decisions. If economics is also put into the equation then, M&S, as an evaluation mechanism, can be utilized for economically rational decisions. With the above in mind, we may, therefore, ask ourselves: "To what extent must we model in order to make our next decision?" and provide the following answer: "To the extent that the corresponding payoff is sufficiently positive" with sufficiently ideally being an exogenous parameter.

2.1. Tradeoffs

When calculating payoff of M&S, a clear view of the tradeoffs is required to set expectations at reasonable levels, conditioned on the risk behavior of choice: aversion/neutrality/love. Two important tradeoffs are fidelity vs. cost and fidelity vs. complexity. Herein, fidelity refers to the quality of the model in terms of describing the observations and predicting the general behavior of the system for process design and operation relevant scenarios. A typical situation is plotted in Figure 2a,b.

Figure 2a provides a visualization of fidelity vs. cost in the plane where a typical progress trajectory is plotted from conception to optimization vs the minimum level of fidelity required for practical applications. We observe that early efforts in a terra incognita result in high cost and low fidelity. Progressively, one reaches the minimum level of fidelity (though with high cost) but further increases in fidelity eventually lead to cost decline and positive cost-effectiveness balance. Consequently, knowledge of where M&S stands with respect to Figure 2a is imperative for accurate calculations of payoffs.

Figure 2b depicts fidelity vs complexity lines; the lines should be perceived as a first-order approximation of the true relation—in reality, complexity vs. fidelity curves have much more complicated structure. The three lines correspond to early, moderate, and mature M&S in a counter-clockwise fashion. The change in slopes denotes how the accumulation of expertise and know-how leads to leaner approaches; for example, via systematic reductions, symmetry considerations, dimensional arguments, clever discretization techniques etc. Similarly to the fidelity vs cost case, the status quo of fidelity vs complexity should be adequately known.

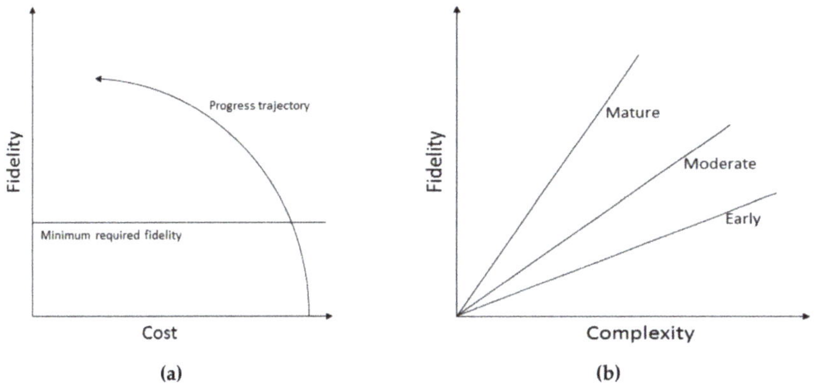

Figure 2. (**a**) Schematic representation of the relationship between fidelity with cost; (**b**) Schematic representation of fidelity vs complexity.

2.2. Decision Flow-Chart

Calculation of payoffs requires an assessment plan of the outcomes of M&S conditioned on the inputs. The flow chart of this process is plotted in Figure 3. We observe that input parameters are cost, time and risk. The output is the results that an M&S exercise yields. It is the assessment of results versus the aggregate effect of cost, time and risk that should drive a go or no go decision for further investment.

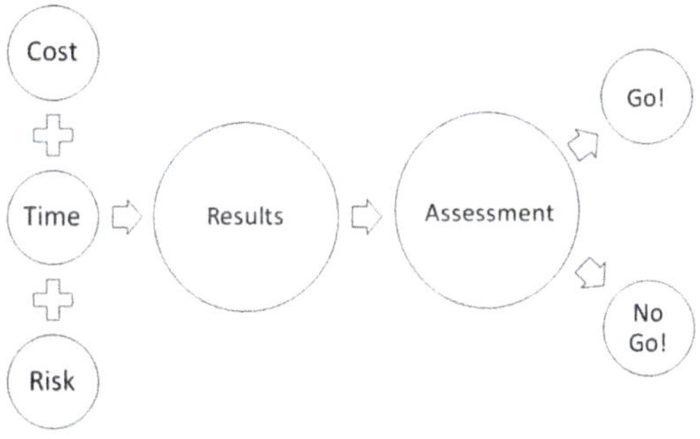

Figure 3. Flow chart of assessment plan for M&S.

For the determination of cost of M&S, we dichotomize models into descriptive and prescriptive ones. Henceforth, A model here is understood as a triplet (governing equations, numerical algorithm/method, software) required for simulation. In other words, it is not only a series of mathematical equations. Descriptive models describe the behavior of existing systems whereas prescriptive models envision to describe the expected behavior of a hypothetical system. For example, a descriptive model would be used to model an existing fermentation vessel. A prescriptive model would be used to design and model a novel fermentation vessel without specific requirements. Descriptive and prescriptive models share similarities with respect to cost, nevertheless, important differences may also be identified. In Table 1, we have tabulated the costs associated with each type of models, further partitioned into upfront (one-time) and recurring costs. One observes the absence of accreditation from prescriptive models; this is to be expected since such models are spearheading R&D and are neither standardized nor subject to systematic upgrades, at least at their early phases of existence. An interesting disparity is the presence of "temptation" in recurring costs. This explains the danger of getting lost in endless exploratory studies, where one goes deeper and deeper whilst there is no clear vision or direction ahead. Cost of temptation may be easy to tame upfront but the lack of a valorization strategy can allow it to skyrocket and undermine budget considerations. Based on Table 1, we can compute the total cost of M&S at year N:

$$Cost(year\ N) = Cost_{soft} + Cost_{hard} + Cost_{train} + Costs_{upgrade} + \#FTEs \cdot Cost_{FTE} \quad (4)$$

Here, $Cost_{soft}$ and $Cost_{hard}$ designated costs related to procurement of software and hardware whereas $Costs_{upgrade}$ stands for maintenance costs and upgrades. $Cost_{FTE}$ is, as usual, the annual cost of a full-time equivalent while $Cost_{train}$ is the cost of related trainings. The concepts of design, implementation, verification, validation, accreditation, and employment are embodied within $Cost_{FTE}$. For the sake of simplicity, we have assumed that FTEs in M&S have the same cost.

Table 1. Cost description for prescriptive and descriptive models.

	Upfront Costs	**Recurring Costs**
Descriptive	Design, Implementation, Verification, Validation, Accreditation, Training, Procurement	Employment, Upgrades
Prescriptive	Design, Implementation, Verification, Validation, Training, Procurement	Employment, Design, Temptation

It is also interesting to visually look at the evolution of cost in time. A linearized picture for both prescriptive and descriptive models is shown in Figure 4. The figure depicts how the temptation point acts as a bifurcation for cost expansion or contraction and how exogenous interventions can act as cost saving mechanisms.

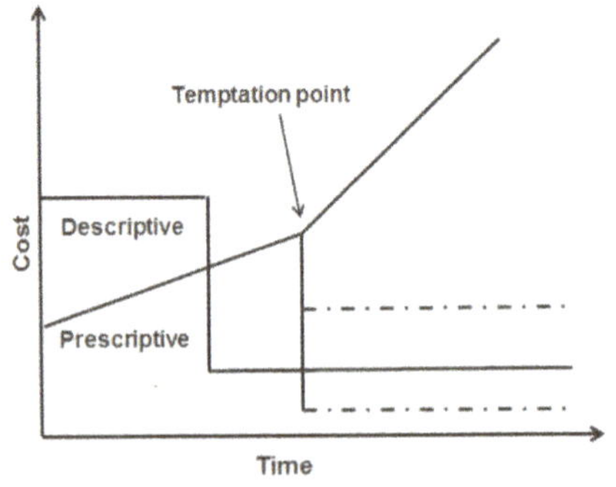

Figure 4. Linearized cost vs. time plots for prescriptive and descriptive models.

Risk associated with M&S has been well documented. In general, M&S risk comprises (i) accuracy, (ii) descriptive realism, (iii) uncertainty and (iv) applicability; each of these components is defined below:

1. Accuracy is defined as the degree to which the predictions are correct (formally, accuracy is defined with respect to a particular norm.).
2. Descriptive realism refers to the degree that a model predicates upon "true" principles [28].
3. Uncertainty refers to the confidence on outputs, given that some aspects are unknown.

4. Applicability accounting for the potential that the exploitation of the model for the envisioned purpose falls short, because the investigated/modeled phenomena do not govern the system in the a priori expected manner.

Each of these components may be viewed as a normalized function that takes values in the range [0, 1]. For accuracy and descriptive realism, zero and unity denote the minimum and maximum values, respectively and the converse is true for uncertainty. We can therefore define the novel aggregate measure of acceptability according to the following formula:

$$Acceptability = \frac{1}{4}(Accuracy + Descriptive\ realism + (1 - uncertainty) + Applicability) \quad (5)$$

where L is labor (FTEs + training) and C is physical capital (hardware/software) and a, b are elasticities with $a + b < 1$. Acceptability takes values in the range [0, 1]; an idea M&S exercise would have 1 accuracy, 1 descriptive and 0 uncertainty thus giving acceptability its maximum value: Unity. (One can go a step further and consider a weighted sum of accuracy, descriptive realism and uncertainty. This preferential aggregate would then reflect a heterogeneous prioritization). For practical purposes, acceptability assumes values in the open range (0, 1). Indeed, even at the start of a modeling effort it is unlikely to expect zero acceptability and reaching unity is typically utopic. We take our analysis a step further and link acceptability to cost. To do so, we consider acceptability as an asset and thus the outcome of a production function. To fix ideas, we postulate that the production function follows a Cobb–Douglas form and thus acceptability obeys the following equation:

$$Acceptability = \frac{1}{4}(Accuracy + Descriptive\ realism + (1 - uncertainty) + Applicability) = L^a C^b \quad (6)$$

where L is labor (FTEs + training) and C is physical capital (hardware/software) and a, b are elasticities with $a + b < 1$.

Proponents of M&S typically invoke time as a competitive advantage. However, the required time for M&S depends on the complexity of the problem in hand and the evolving technology and know-how. Figure 5 portrays the trends of time versus complexity for the past decades. As expected, evolution in hardware/software and physical modeling itself pushes the curve in a southeast direction. However, despite this rather robust shift, time remains an exponential function of complexity.

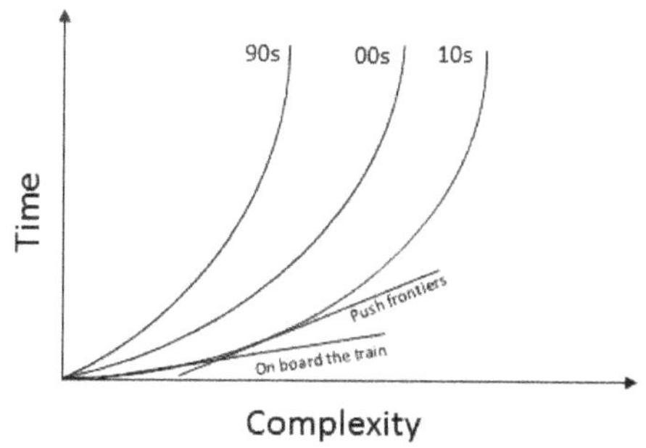

Figure 5. Qualitative assessment of required time for M&S. The numbers represent the decades.

Mathematically, progress increases the part of the curve that can be accurately linearized. This domain is labeled "on board the train" to emphasize that in this area one takes advantage of the accumulated advancements. It is in this zone where time and complexity correlate in a favorable manner. On the right of the "on board the train zone" is the "push the frontiers zone" where a linearized curve changes slope and no longer provides an accurate fit. Here is where innovation mostly occurs. Risk aversion dictates the avoidance of the purely exponential region and the focus on the "push the frontiers" one.

We can further quantify the time required for M&S. Our starting point is the observation that aggregate time required for a M&S exercise can be written as the following sum:

$$T_{total} = T_{model} + T_{digitization} + T_{simulation} + T_{interpretation} \quad (7)$$

where T_{model}, $T_{digitization}$, $T_{simulation}$, $T_{interpretation}$ designate the time needed for the development of a model, digitization of equipment, numerical simulation, post-processing and interpretation. An order-of-magnitude analysis can be used to provide some estimates which are reported in Table 2.

Table 2. Order-of-magnitude analysis of required time for M&S per component.

Time Required	T_{model}	$T_{digitization}$	$T_{simulation}$	$T_{interpretation}$
~DAYS	Reuse	Reuse	Low & medium complexity	No post-processing required
~WEEKS	Reuse/discover	Digitize existing system	Detailed CFD	Meticulous post-processing/big data
~MONTHS	Develop	Design & digitize system	Industrial scale CFD	N/A

We remark that in the case of first time used or newly developed models, T_{model} accommodates the validation phase as well. Experience has shown that, within pharma, this phase can be quite elongated, as it often involves a chain of actors from non-scientific departments. Thus, one should not underestimate such exogenous factors when drafting (or predicting) time schedules.

3. An Investor's Approach to M&S

With all the above in mind, we can again call upon our investor, budget owner Mr. X who, correctly or not, has already invested in a M&S team for some time now, being aware of the uncertainty that dominates this decision. Mr. X has now to harvest the results of the investment and needs to define a payoff measure. For a quantitative assessment, Mr. X has to attribute a set of relevant metrics to the corresponding outcomes. This set of metrics will be decomposed into monetary metrics and performance metrics. The need (or rationale) behind this decomposition is as follows. The overall investment has had but a short life and aims in implicitly increasing the net revenue per employee by enhancing the competences that employees have at their disposal. This is not an instantaneous process as Figure 1b asserts. In this respect, Mr. X should keep track not only of monetary payoffs but also of intangible metrics that evaluate the integration of M&S alongside existing R&D practices.

3.1. Monetary Metrics

The first bottleneck is the identification of gains or equivalent the payoff. Three monetary metrics appear as the most prominent candidates: Cost savings, cost avoidance and increased revenues. Formally, they are defined as follows:

1. Cost savings = Cost with M&S—Cost without M&S
2. Cost avoidance = Cost of unnecessary/harmful decision.
3. Increased revenues = profit due to changes in margins or production capacity.

Each of the above monetary metrics can have a single or permanent impact on the sector. Cost savings has a single impact because it refers to gains that do not affect permanently the production capacity and/or revenue. For instance, they may refer to cost savings in a project that failed and never reached production. Cost avoidance has also single impact. It concerns multi-lemmas that once resolved it is for permanent; for example, consider the case where a company needs to decide in favor of one type of instrument vs another. Finally, increased revenues have permanent impact in the corporation. This is a result of M&S permanently affecting the profit margin. (Calculating the above metrics in practice is easier said than done and the typical example is that of knowledge-build projects).

Having collected the required data, Mr. X proceeds to compute a time-dependent return on investment as follows. Consider a time interval $[t_0, t_n]$ uniformly discretized into time instances t_i such that $t_i - t_{i-1} = \Delta t = constant$. Δt can assume any value, e.g., a month. For each t_i calculate the M&S cashflow:

$$\begin{aligned} C^{MS}(t_i) &= Cost\ savings(t_i - t_{i-1}) + Cost\ avoidance(t_i - t_{i-1}) \\ &+ Increased\ revenues(t_i - t_{i-1}) - Investment\ cost(t_i - t_{i-1}) \end{aligned} \quad (8)$$

$Cost\ savings(t_i - t_{i-1})$ denotes the costs savings during the period $t_i - t_{i-1}$ and the same applies for the other components. Consider that $C^{MS}(t_i)$ have been measured from $t_0 = 0$ until present $t_{present} = K_{present}\Delta t$. Then, let $N_{proj} \geq 1$ denote the number of projects that M&S personnel have been working on during this time. Each project has an anticipated duration of $T_j = K_j \Delta t$, with $j = 1, \ldots, N_{proj}$ and $T_j > t_{present}$ $\forall j$ and is expected to induce an internal rate of return (IRR) IRR_j, which is a solution to the following;

$$NPV = \sum_{i=0}^{K_j} \frac{E(C^{tot}{}_j(t_i))}{\left(1 + IRR_j\right)^i} = 0 \quad (9)$$

and obeys the following inequality:

$$IRR_j > RRR \quad (10)$$

In the above relations, $E\left(C_j^{tot}(t_i)\right)$ stands for the expectation of the total cash-flow of project j at time instance t_i while RRR is the required rate of return which stands for the minimum accepted rate that renders the investment rationally possible. (This is a rather traditional approach. Alternative methodologies that use the real option value such as the Datar–Mathews method, and incorporate risk, may be also utilized) [29].

Next, Mr. X applies a Bayesian inference of the time-series $C_{MS}(t_i)$, as duly demonstrated in Figure 6, and calculates stochastic predictions for the evolution of cashflows from $t = t_{present} + \Delta t$ until $T_{max} = \max\{T_j\} = K_{max}\Delta t$. Thus, Mr. X obtains a spatio-temporal probability distribution that assigns to each possible $C_{MS}(t_i)$ a probability. Then, probabilistic estimates of the IRR_{MS} of the M&S investment can be computed according to;

$$NPV_{MS} = \sum_{i=0}^{K_{max}} \frac{C_{MS}(t_i)}{(1 + IRR_{MS})^i} = 0,\ P(IRR_{MS}) = p \quad (11)$$

for all possible outcomes enveloped by the Bayesian inference and a graph, like the one reported in Figure 7 displaying predicted IRR_{MS} against probabilities, can be constructed.

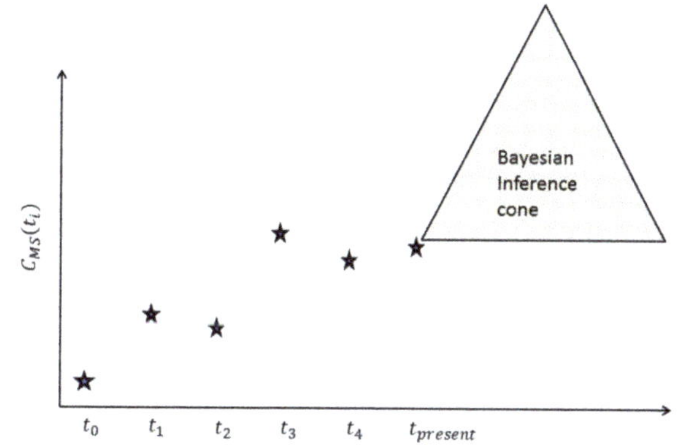

Figure 6. Hypothetical data points of $C_{MS}(t_i)$ vs. time, shown with $*$, concatenated with Bayesian inference.

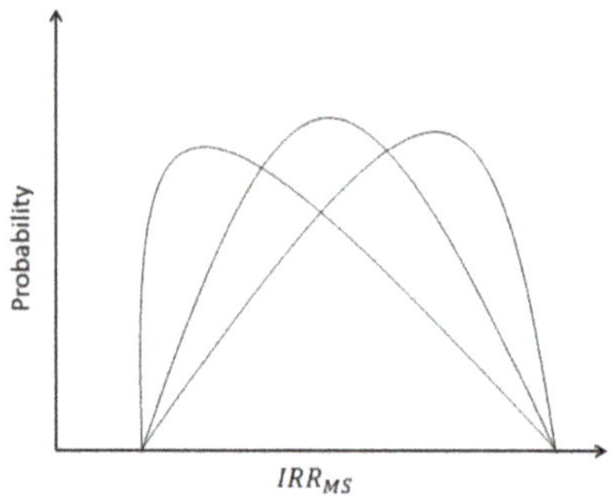

Figure 7. Predicted IRR_{MS} values vs their probability for three hypothetical cases resulting in three different probability distributions.

Next, a lower limit of acceptance (LLA) for the investment is developed. Assume that $IRR_{MS} \sim N(\mu, \sigma^2)$. Ostensibly, one could equate LLA with the expected rate of returns RRR. However, we propose a modification so that the degree of diffusion of M&S within the organization is captured. The diffusion of a new technology can be satisfactorily described via a sigmoid function $(t) = \frac{1}{1+e^{-\delta t}}$. (The parameter δ determines the time needed for M&S to completely diffuse, i.e., at which point where the function equals unity. It can be estimated with the help of the intangible metrics of the next section. Other more elaborate and asymmetric functions may also be considered.). Then, a reasonable lower limit of acceptance at time t_i is as follows:

$$LLA(t_i) = S(t_i)RRR \tag{12}$$

We can reconcile the stochastic nature of IRR_{MS} with the risk aversion of Mr. X and Equation (12) as follows. We assign to Mr. X a CARA (constant absolute risk aversion) utility function $u(x) = 1 - e^{-Ax}$ with A being the known risk aversion coefficient, that can be estimated via the methodology of [30], and we recall the budget of $M\$$. Consider the portfolio allocation problem with one risky asset (investment on M&S) with random return IRR_{MS} and a riskless asset with fixed return $LLA_{MS}(T_{max})$. Under rationality, we can explicitly solve this two-asset (one risky and one riskless) portfolio allocation problem, see for example [31]. This solution asserts that that recourses should be allocated if and only if;

$$\mu > LLA_{MS}(T_{max}) \tag{13}$$

and that the optimal degree of allocation obeys the following condition:

$$M_{MS} = \frac{\mu - LLA_{MS}(T_{max})}{\sigma^2} \cdot \frac{1}{A} \tag{14}$$

Equations (13) and (14) have practical implications. First, Mr. X determines whether (13) is satisfied. If yes, then given $M_1\$$ have already been invested in M&S, Mr. X can solve (14) in terms of the expected return $\mu\prime$, run a sensitivity analysis for σ^2 and end up with a range of values $\left[\mu_{low}, \mu_{high}\right]$. Then, Mr. X can compare how well the predictions compare to their realizations, or equivalently where μ lies in the range $\left[\mu_{low}, \mu_{high}\right]$.

This is a decision tree with two negative outcomes. First, the realized expected return μ is lower than the lower acceptable limit, i.e., inequality (13) is violated. Thus, the overall investment rates are unfavorable. As the distribution of IRR_{MS} is calculated based on Bayesian inference, it is updated as soon as new data enter the system. Therefore, one should re-evaluate the overall investment at time instances $t_i > t_{present}$ and check if violation of inequality (13) is an artifact or not (the generation of monetary gains might come with a (random) time delay). The second negative outcome concerns Equation (14) and the range $\left[\mu_{low}, \mu_{high}\right]$. If $\mu < \mu_{low}$ then the investment in M&S is still worthwhile but the expected return is probably overestimated. By contrast, if $\mu > \mu_{high}$ then the investment outperforms expectations.

3.2. Diffusion-of-Innovation Metrics

Diffusion-of-innovation metrics help our investor, Mr. X, to make better informed decisions when making assessments. The authors of [32] proposed a notoriously high (over 200) number of metrics that constitute assessment criteria of how well modeling and simulation is deployed within the US Department of Defense. Therefore, to a first approximation, this pool of metrics can used for the selection of diffusion-of-innovation metrics that pertain to our case. Of course, it is not only the impractically large number that renders our task challenging but also the fact that several of these metrics are bespoke to the army needs. By merging and redefining available metrics, so as to fit the pharmaceutical world, we have arrived at the following seven (7) performance related metrics, delineated in Table 3 alongside their numerical value.

Table 3. Diffusion metrics and their numerical value.

Name of Metric	Numerical Value (s)
Awareness	Relative frequency of different projects utilizing M&S
Coordination	Relative frequency of M&S duplicate activities avoided
Congruity	Relative frequency of M&S clients correctly interpreting/understanding the results
Guidance	Relative frequency of M&S users conforming to existing standards
Proactivity	Relative frequency of (early) decisions made by M&S
Empowerment	Relative frequency of M&S decision makers attending key meetings
Foundation	Relative frequency of foundational competencies covered

As the numerical values increase and approach unity so does the integration of M&S in R&D. The numerical values of the metrics s_i, $i = 1, \ldots, 7$ are functions of time, i.e., $s_i = s_i\left(t_j\right)$ where the values t_j conform to the previous section discussion. Next, consider the sigmoid function $\frac{1}{1+e^{-\delta t}}$ which has the diffusion rate as a free parameter, δ. Also, consider the sum $\frac{1}{7}\sum_{j=0}^{K_{present}} s_i\left(t_j\right)$. If the sum of the metrics provides a satisfactory description of the diffusion of M&S in R&D, then it is reasonable to consider the following approximation:

$$\frac{1}{1+e^{-\delta t}} \approx \frac{1}{7}\sum_{j=0}^{K_{present}} s_i\left(t_j\right) + u \tag{15}$$

where $u = N\left(0, \sigma_u^2\right)$ is a white noise term to reflect the fact that the diffusion process is associated with a certain degree of randomness and can be amenable to random shocks (depending on the strategy that leadership has developed, the diffusion metrics could be assigned a weight $\left(t_j\right)$ with $w_1\left(t_j\right) + w_2\left(t_j\right) + \ldots + w_6\left(t_j\right) = 1$; note that the weights are also functions of time to reflect reprioritizations and changes in strategy). Then, the constant δ may be estimated via simple regression from the above equation and directly utilized in Equation (12) of the previous section.

4. Conclusions

Process modeling is gradually gaining momentum within the pharmaceutical industry. This inevitably attracts attention from higher management and onsets the discussion of cost-benefit analysis and investment decisions. This paper has examined process modeling from the investor's point of view.

We have commenced by examining whether current practices conform to a value-based decision process by using an informed investor as the decision maker. Further, topics like cost, risk and execution time for M&S exercises have also been thoroughly discussed and, wherever possible, mathematical expressions for their description have been introduced. We subsequently proceeded to the development of an easy-to-use methodology that can help an investor evaluate investment on M&S that encompasses both monetary and diffusion-of-innovation based aspects. The proposed methodology builds upon a classical discounted cash flow analysis, infused with elements of Bayesian inference, while accommodating a sigmoid-description of diffusion of innovation for the calculation of lower expected returns. Via the introduction of a set of seven intangible metrics we were also able to quantify the rate with which M&S diffuses within the organization thereby rendering the proposed methodology tractable.

In the present study, we have limited ourselves to the theoretical presentation of the mathematical model were emphasis has been placed on clarifying ideas and concepts. A natural next steps involves exercising the proposed model with either real or simulated data and ideally with both. This step that we intend to pursue as a follow up to this work will play a pivotal role in assessing the predictive capacity of our methodological framework and underlying possible weaknesses that need mitigation.

This concept paper comprises different components that can collectively assist with the difficult task of evaluating M&S. However, herein, we have restricted ourselves to the illustration of these ideas without emphasis on their connectedness. Consequently, a further direction of future research concerns the unification of all ideas presented herein in an umbrella framework.

Author Contributions: C.V., M.v.S., S.D. and A.P. were involved in the design of the study. All authors were involved in drafting the manuscript or critically revising it for important intellectual content. All authors approved the manuscript before it was submitted by the corresponding author.

Funding: This research received no external funding.

Acknowledgments: The authors would like to thank the anonymous reviewers for valuable remarks and suggestions that have substantially improved the quality of the manuscript. Moreover, we thank S. J. E. Evans, N. Giannelos, A. Khan, U. Krause, G. De Lannoy, M. Sanders and M. Vasselle for fruitful discussions and remarks.

Conflicts of Interest: All authors have declared the following interests: All authors are employees of the GSK group of companies. S.D. and A.P. report ownership of shares and/or restricted shares in the GSK group of companies.

References

1. Petrides, D.P.; Koulouris, A.; Lagonikos, P.T. The Role of Process Simulation in Pharmaceutical Process Development and Product Commercialization. *Pharm. Eng.* **2002**, *22*, 56–65.
2. García-Muñoz, S.; Luciani, C.V.; Vaidyaraman, S.; Seibert, K.D. Definition of Design Spaces Using Mechanistic Models and Geometric Projections of Probability Maps. *Org. Process Res. Dev.* **2015**, *19*, 1012–1023. [CrossRef]
3. Aboud, L.; Henry, S. New Prescription for Drug Makers: Update the Plants. Leila Aboud & Scott Henry. *The Wall Street Journal*. 3 September 2003. Available online: https://www.wsj.com/articles/SB106253584039331000 (accessed on 4 September 2019).
4. Rogers, A.; Ierapetritou, M. Challenges and opportunities in modeling pharmaceutical manufacturing processes. *Comput. Chem. Eng.* **2015**, *81*, 32–39. [CrossRef]
5. Muzzio, F.J.; Shinbrot, T.; Glasser, B.J. Powder technology in the pharmaceutical industry: The need to catch up fast. *Powder Technol.* **2002**, *124*, 1–7. [CrossRef]
6. McKenzie, P.; Kiang, S.; Tom, J.; Rubin, A.; Futran, M. Can pharmaceutical process development become high tech? *AIChE J.* **2006**, *52*, 3990–3994. [CrossRef]
7. Reklaitis, G.V.; Khinast, J.; Muzzio, F. Pharmaceutical engineering science—New approaches to pharmaceutical development and manufacturing. *Chem. Eng. Sci.* **2010**, *65*, 4–8. [CrossRef]
8. Eberle, L.G.; Sugiyama, H.; Papadokonstantakis, S.; Graser, A.; Schmidt, R.; Hungerbühler, K. Data-driven Tiered Procedure for Enhancing Yield in Drug Product Manufacturing. *Comput. Chem. Eng.* **2016**, *87*, 82–94. [CrossRef]
9. Casola, G.; Siegmund, C.; Mattern, M.; Sugiyama, H. Uncertainty-conscious methodology for process performance assessment in biopharmaceutical drug product manufacturing. *AIChE J.* **2018**, *64*, 1272–1284. [CrossRef]
10. Van Bockstal, P.J.; Mortier, S.; De Meyer, L.; Corver, J.; Vervaet, C.; Nopens, I.; De Beer, T. Mechanistic modelling of infrared mediated energy transfer during the primary drying step of a continuous freeze-drying process. *Eur. J. Pharm. Biopharm.* **2017**, *114*, 11–21. [CrossRef]

11. Kornecki, M.; Strube, J. Accelerating Biologics Manufacturing by Upstream Process Modelling. *Processes* **2019**, *7*, 166. [CrossRef]
12. Metta, N.; Ghijs, M.; Schäfer, E.; Kumar, A.; Cappuyns, P.; Van Assche, I.; Singh, R.; Ramachandran, R.; De Beer, T.; Ierapetritou, M.; et al. Dynamic Flowsheet Model Development and Sensitivity Analysis of a Continuous Pharmaceutical Tablet Manufacturing Process Using the Wet Granulation Route. *Processes* **2019**, *7*, 234. [CrossRef]
13. Chatterjee, S.; Moore, C.; Nasr, M. An Overview of the Role of Mathematical Models in Implementation of Quality by Design Paradigm for Drug Development and Manufacture. *Food Drug Adm. Papers* **2017**, 23.
14. Matsunami, K.; Miyano, T.; Arai, H.; Nakagawa, H.; Hirao, M.; Sugiyama, H. Decision support method for the choice between batch and continuous technologies in solid drug product manufacturing. *Ind. Eng. Chem. Res.* **2018**, *57*, 9798–9809. [CrossRef]
15. Rantanen, J.; Khinast, J. The Future of Pharmaceutical Manufacturing Sciences. *J. Pharm. Sci.* **2005**, *104*, 3612–3638. [CrossRef]
16. Gernaey, K.V.; Woodley, J.; Sin, S. Introducing mechanistic models in Process Analytical Technology education (Research Highlight). *Biotechnol. J.* **2009**, *4*, 593–599. [CrossRef]
17. Deloitte Center for Health Solutions. *A New Future for R&D? Measuring the Return from Pharmaceutical Innovation*; Deloitte Centre for Health Solutions: Deloitte, UK, 2017.
18. DiMasi, J.A.; Hansen, R.W.; Grabowski, H.G. The price of innovation: New estimates of drug development costs. *J. Health Econ.* **2003**, *22*, 151–185. [CrossRef]
19. Grabowski, H.; Vernon, J.J. A new look at the returns and risks to pharmaceutical R&D. *Manag. Sci.* **1990**, *36*, 804–821.
20. David, E.; Tramontin, T.; Zemmel, R. Pharmaceutical R&D: The road to positive returns. *Nat. Rev. Drug Discov.* **2009**, *8*, 609–610.
21. Carter, J., III. *A Business Case for Modeling and Simulation*; SPECIAL REPORT-RD-AS-01-02; Aviation and Missile Research, Development, and Engineering Center: Redstone Arsenal, AL, USA, 2001.
22. Oswalt, I.; Cooley, T.; Waite, W.; Waite, E.; Gordon, S.; Severinghaus, R.; Feinberg, J.; Lightner, G. *Calculating Return on Investment for U.S. Department of Defense Modeling and Simulation*; Defense Acquisition Univ. ft. Belvoir VA: Fort Belvoir, VA, USA, 2011.
23. Brown, D.; Grant, G.; Kotchman, D.; Reyenga, R.; Szanto, T. Building a business case for modeling and simulation. *Acquis. Rev. Q.* **2000**, *24*, 312–315.
24. Smith, J.M. *A Business Case for Using Modeling and Simulation in Developmental Testing*; Thesis Naval Postgraduate School; Storming Media: Washington, DC, USA, 2001.
25. Gordon, S.; Oswald, I.; Cooley, T. Why Spend One More Dollar for M&S? Observations on the Return of Investment: Discipline, Ethics, Education, Vocation, Societies, and Economics. In *The Profession of Modeling and Simulation*; Chapter 14; John Wiley & Sons: Hoboken, NJ, USA, 2017.
26. Glass, H.E.; Kolassa, E.M.; Muniz, E. Drug development through modeling and simulation-The business case. *Applied Clinical Trials*. 25 July 2016. Available online: http://www.appliedclinicaltrialsonline.com/drug-development-through-modeling-and-simulation-business-case (accessed on 4 September 2019).
27. Rogers, E. *Diffusion of Innovations*, 5th ed.; Simon and Schuster: New York, NY, USA, 2003.
28. Meyer, W.J. *Concepts of Mathematical Modeling*; McGraw-Hill Book Company: New York, NY, USA, 1984.
29. Mathews, S.H.; Datar, V.T.; Johnson, B. A practical method for valuing real options. *J. Appl. Corp. Financ.* **2007**, *19*, 95–104. [CrossRef]
30. Babcock, B.A.; Choi, E.K.; Feinerman, E. Risk and probability premiums for CARA utility functions. *J. Agric. Resour. Econ.* **1993**, *18*, 17–24.
31. Back, K.E. *Asset Pricing and Portfolio Choice Theory*; Oxford Univeristy Press: Oxford, UK, 2017.
32. AEgis Technologies Group. *Metrics for Modeling and Simulation (M&S) Investments*; Naval Air Systems Command Prime Contract No. N61339-05-C-0088; The Aegis Technologies Group, Inc.: Huntsville, AL, USA, 2008.

MDPI
St. Alban-Anlage 66
4052 Basel
Switzerland
Tel. +41 61 683 77 34
Fax +41 61 302 89 18
www.mdpi.com

Processes Editorial Office
E-mail: processes@mdpi.com
www.mdpi.com/journal/processes

www.ingramcontent.com/pod-product-compliance
Lightning Source LLC
LaVergne TN
LVHW071949080526
838202LV00064B/6710